Digital Transformations in the Challenge of Activity and Work

Technological Changes and Human Resources Set

coordinated by
Patrick Gilbert

Volume 3

Digital Transformations in the Challenge of Activity and Work

Understanding and Supporting Technological Changes

Edited by

Marc-Eric Bobillier Chaumon

iSTE

WILEY

First published 2021 in Great Britain and the United States by ISTE Ltd and John Wiley & Sons, Inc.

ISTE Ltd
27-37 St George's Road
London SW19 4EU
UK

www.iste.co.uk

John Wiley & Sons, Inc.
111 River Street
Hoboken, NJ 07030
USA

www.wiley.com

Library of Congress Control Number: 2020946633

British Library Cataloguing-in-Publication Data
A CIP record for this book is available from the British Library
ISBN 978-1-78630-529-9

Contents

Introduction xiii

Marc-Eric BOBILLIER CHAUMON

Part 1. Towards an Overview of Digital Transformations... 1

Chapter 1. Emerging Technologies and Issues for Activity and Occupational Health. 3

Marc-Eric BOBILLIER CHAUMON

1.1. Introduction 3
1.2. From properties to the uses of emerging technologies 4
1.3. Five paradoxes of the diffusion of technologies in/on the activity 8
1.3.1. Sense of loss of control over the activity vs increased control over the activity 8
1.3.2. Invisibility vs visibility of the activity 9
1.3.3. Increase in virtual teams vs isolation of employees 10
1.3.4. Nomadism vs a sedentary lifestyle at work. 11
1.3.5. Detachment from activity vs proximity of work 12
1.4. Conclusion 14
1.5. References 15

Chapter 2. Collaborative Work Platforms: Challenges for Business Development 21

Nadia BARVILLE-DEROMAS and Marc-Eric BOBILLIER CHAUMON

2.1. Introduction 21
2.2. Two organizational challenges: empowering digital transformations and changing work practices. 22
2.2.1. The co-configuration of work 22
2.2.2. Mobilizing social capital to make sharing more flexible 23

2.3. Stakes for the development of activity: knowing how to give meaning to a poly-contextual and multi-mediated activity 24
2.3.1. Networking, the power to act and meaning at work 24
2.3.2. Sharing a degree of collaborative intentionality in a multi-mediated situation, a skill in its own right. 26
2.4. Conclusion 27
2.5. References 27

Chapter 3. Virtual Reality: Definitions, Characteristics and Applications in the Workplace 31

Camille SAGNIER, Émilie LOUP-ESCANDE and Gérard VALLÉRY

3.1. Introduction 31
3.2. Some elements of definition 32
3.2.1. The term "virtual reality". 32
3.2.2. The purpose of virtual reality 32
3.2.3. A functional definition of virtual reality 33
3.2.4. A technical definition of virtual reality 33
3.3. The main interaction devices 34
3.3.1. Display devices 35
3.3.2. Motion and position capture devices 35
3.3.3. Proprioceptive and cutaneous feedback devices 35
3.3.4. Sound input and presentation devices. 36
3.4. The main areas of application of virtual reality 36
3.4.1. Applications in everyday life 37
3.4.2. Applications in various professional contexts 37
3.5. Applications of virtual reality in industry 38
3.6. Conclusion 40
3.7. References 40

Chapter 4. Robotization in Industries: A Focus on SMEs 45

Sandrine BERGER-DOUCE

4.1. Introduction 45
4.2. Focus on a robotization experiment in an industrial SME 46
4.2.1. The case study in the service of exploratory research 46
4.2.2. Presentation of the company. 47
4.2.3. A mixed technological adventure 47
4.2.4. Lessons to be learned. 48
4.3. Receiving support in order to better implement a robot: illustration by the Robot Start PME program 49

4.3.1. A program at the service of French industrial SMEs 50
4.3.2. An inspiring framework to support technological change 53
4.4. Conclusion 54
4.5. References 55

Chapter 5. Serious Games for Vocational Training: From Emotional Labor to Knowledge Transfer 57

Lydia MARTIN, Julian ALVAREZ and Antoine TALY

5.1. Introduction 57
5.2. Emotions, debriefing and learning 59
5.2.1. Emotional labor 59
5.2.2. Learning in a game 60
5.3. The context and framework of the Serious Escape Game (SEG) 61
5.3.1. Difficulty in gamification 61
5.3.2. The training system 62
5.3.3. A research-action approach 62
5.4. Results 63
5.4.1. Observations during gaming 63
5.4.2. Post-game discussions 64
5.4.3. Questionnaires. 66
5.5. Discussion and conclusion 67
5.6. References 68

Chapter 6. The "Old" Issues of the "New" Artificial Intelligence Systems in Professional Activities 71

Tamari GAMKRELIDZE, Moustafa ZOUINAR and Flore BARCELLINI

6.1. Introduction 71
6.2. AI: elements of definition and recent developments 72
6.3. Functionalities and (potential) uses of new generation AI systems 74
6.4. The "new" generation of AI and the old challenges of transforming work situations 77
6.4.1. From the substitutive approach of automation to the complementary approaches of human–machine systems 77
6.4.2. The challenges of "new" AI systems in work situations. 79
6.5. What are the approaches to designing and integrating AI systems in work situations?. 81
6.6. Conclusion 83
6.7. References 83

Part 2. New Modalities and Forms of Work... 87

Chapter 7. Challenges in Deploying Telework: Benefits and Risks for Employees 89

Emilie VAYRE

7.1. Telework: definitions and characteristics 89
7.2. The benefits of teleworking 90
7.3. The constraints and risks of teleworking 91
7.4. The challenges of deploying telework in organizations 93
7.4.1. Deploying and experimenting with telework 93
7.4.2. Training of teleworkers and managers 96
7.4.3. Evaluating the deployment of telework 97
7.5. Conclusion 97
7.6. References 98

Chapter 8. The Reconfiguration of Managerial Practices through Digital Innovation: The Example of a Work Team in Site Renovation 101

Elodie CHAMBONNIÈRE and Jacqueline VACHERAND-REVEL

8.1. Introduction: when digital technology is used on renovation sites 101
8.2. At the heart of the renovation sites 103
8.2.1. Supervising in a complex and dynamic system 103
8.2.2. Guiding a worksite: a conductor's activity at the crossroads of various modes of prevention management 103
8.3. Understanding occupational risk prevention activity and prevention management 104
8.4. Ethnography of the activity on a renovation site 106
8.5. Confirming a culture of safety: prevention management 107
8.5.1. Management towards site supervision 108
8.5.2. Middle management 109
8.5.3. Local management: towards construction workers 109
8.6. Digital innovation in occupational risk prevention: restructuring of management practices 110
8.6.1. Hierarchical visits by management 110
8.6.2. Prevention visits by middle management 111
8.6.3. Close supervision of the construction workers 111
8.7. Conclusion: towards a better consideration of digital innovations in prevention management 113
8.8. References 114

Chapter 9. Integrating Collaborative Robotics into Work Situations: The Intentions of SME Managers in the Digital Transformation of their Companies . . . 115

Anne-Cécile LAFEUILLADE, Flore BARCELLINI, Willy BUCHMANN and Tahar-Hakim BENCHEKROUN

9.1. Transformations in work situations seen through the prism of technocentric solutions . . . 115
9.2. Models of leadership activity to understand change management processes . . . 117
9.2.1. The activity of managers at the crossroads of different roles . . . 117
9.2.2. Developing the intention of managers in change management processes: the contribution of the dialogical model of design . . . 118
9.3. Methodology for data collection and analysis . . . 121
9.4. Managers' desires in the face of reality: an encounter that helped to shape their intentions . . . 123
9.4.1. Elements shaping managers' desires . . . 123
9.4.2. The "conversation" between the desire and reality . . . 124
9.5. The reality, a messenger from the past, in a modernization project . . . 125
9.6. References . . . 126

Chapter 10. The Role and Function of Technological Artifacts in Entrepreneurial Activity . . . 129

Irène POIDI, Marc-Eric BOBILLIER CHAUMON and Jacqueline VACHERAND-REVEL

10.1. Introduction . . . 129
10.2. Theoretical foundations . . . 130
10.3. Methodology . . . 132
10.4. Results . . . 133
10.5. Discussion and conclusion . . . 137
10.6. References . . . 138

Part 3. Psychosocial and Socio-organizational Impacts of the Diffusion of Technology . . . 141

Chapter 11. The New Physical Territories of Digital Activity . . . 143

Maria IANEVA, Raluca CIOBANU and Chiara LAI

11.1. Introduction . . . 143
11.2. Transformation of spaces and transformation of work and employment: "spatialized work" . . . 145
11.3. From "spatialized work" to the division between space and work . . . 146
11.4. Flexible work environments: from work to "activity" . . . 147

11.4.1. The example of the design of the workspaces of a large company: the reconfiguration of work areas . . . 148
11.4.2. From space allocation to the redefinition of associated tasks . . . 149
11.5. What theoretical models for considering space and its transformations? . . . 150
11.6. Conclusion . . . 152
11.7. References . . . 153

Chapter 12. Digital Work, Disposable Work? When Opportunities to Explore Threaten the Meaning of the Activity . . . 155

Maxime BESENVAL and Alexandra BIDET

12.1. Introduction . . . 155
12.2. The division of complex digital work . . . 157
12.3. Chronic indeterminacy of the product . . . 159
12.4. When the contingencies regime threatens work commitment . . . 161
12.5. Conclusion . . . 165
12.6. References . . . 166

Chapter 13. Is the Obsolescence of the Skills of Older Employees an Inevitable Consequence of Digitalization? . . . 169

Florence CROS, Marc-Eric BOBILLIER CHAUMON and Bruno CUVILLIER

13.1. Introduction . . . 169
13.2. Aging, work, technologies and skills obsolescence: theoretical elements . . . 170
13.2.1. The effects of aging . . . 170
13.2.2. Aging associated with work . . . 171
13.2.3. From the effects of technology on activity to the issue of the obsolescence of older workers' skills . . . 171
13.3. Question and methodology . . . 172
13.3.1. Study background and methods . . . 172
13.3.2. Engeström's activity system model (1987) . . . 173
13.4. Main results . . . 174
13.4.1. The ASCT profession: a perception through the prism of age . . . 174
13.4.2. Overcoming tensions to develop one's activity . . . 176
13.5. Discussion and conclusion . . . 177
13.5.1. Accelio, a vector of recognition for the ASCT profession? . . . 178
13.5.2. Diverting to work better . . . 178
13.6. References . . . 179

Chapter 14. Are Work Collectives and Digital Exposure Compatible? . . . 183

Pauline CROUZAT and Marc-Eric BOBILLIER CHAUMON

14.1. Collective activity: major developments 183
14.2. Engineering: a highly digitized environment 185
14.3. Problem of the study 185
14.4. The methodology used 186
14.5. Main results: virtually hyper-instrumented collective work for invisible work collectives 190
14.6. Discussion. 191
14.7. Conclusion 192
14.8. References 193

Part 4. Approaches and Methods for Conducting Digital Transformations 195

Chapter 15. Prospective Ergonomics in Service of Technological Innovation 197

Jean-Marc ROBERT, Antoine MARTIN, Mitra TARAGHI, Clément COLIN, Masood MALDAR, Flavie BONNEVIOT and Eric BRANGIER

15.1. Introduction. 197
15.2. A new form of intervention focused on creation and innovation. 198
15.3. The context and challenges of prospective ergonomics. 199
15.4. Foundations in several disciplines and fields of study 202
15.5. Prospective ergonomics intervention approach 203
15.6. Two cases of intervention 206
15.6.1. Anticipating future needs and uses related to energy systems that integrate hydrogen energy 206
15.6.2. The crowdsourcing approach. 207
15.7. Conclusion 208
15.8. References 209

Chapter 16. Simulating Digital Activity in the Making: Elements of Methodology 211

Laurent VAN BELLEGHEM

16.1. Introduction. 211
16.2. Digital activity 212
16.3. Simulating digital activity 213
16.4. Two illustrations of digital activity simulation 217
16.4.1. A giant digital tablet to welcome customers 217
16.4.2. RFID chips for lending books 219

16.5. Conclusion 222
16.6. References 223

Chapter 17. Managing Technological Change 225

Clotilde CORON and Patrick GILBERT

17.1. Introduction 225
17.2. Digital transformations, sources of threats and opportunities 226
17.3. Social and responsible management of new technologies 229
17.3.1. Social and responsible innovations 230
17.3.2. Digital and responsible technological innovation 230
17.4. A model for responsible leadership of technological change within organizations 232
17.4.1. Managing organizational and technological change 232
17.4.2. Responsible digital change management 233
17.5. Conclusion 234
17.6. References 235

Chapter 18. Exploring the Situated Acceptance of Emerging Technologies in and Concerning Activity: Approaches and Processes 237

Marc-Eric BOBILLIER CHAUMON

18.1. Introduction 237
18.2. Models of technological acceptability: outlines and approaches 239
18.2.1. Practical or instrumental acceptability 239
18.2.2. The social acceptability model 240
18.2.3. The situated acceptance approach 242
18.3. Frameworks for action and intervention to address situated acceptance 244
18.3.1. Evaluating the situated acceptance of technologies already in place to support the future technological transformation project 246
18.3.2. Exploring the acceptance of emerging technologies to co-construct the creation project in/by the activity 247
18.4. Conclusion 252
18.5. References 252

List of Authors 257

Index 261

Introduction

There has been an acceleration in technological change within the workplace. This is because innovations and the deployment of technical solutions are multiplying (*AI*, *robotization*, *immersive environment*, *Big Data*, *collaborative platforms*, *factory of the future*, etc.). At best, the implementation of these new devices provokes simple transfers of use and learning (which are iterative or incremental technologies), and, at worst, they provoke real disruptions in use (which are disruptive technologies) (Bobillier Chaumon 2016). These can require a profound rethinking of the user experience, which can cause employees to fail and make it more difficult for them to carry out their activity. Therefore, new organizational and socio-cognitive models are required, as the devices call for alternative ways of thinking, doing and collaborating at work.

Professional activity thus depends less and less on the direct intervention of individuals in terms of the object of work than on their actions within these digital work environments, that is, on technological artifacts that mediate this activity. This distancing phenomenon is consubstantial with the quality of the "human–machine" systems used: if a system is complex to use (i.e. designed with an insufficient level of usability) or if it is difficult to attribute meaning to this technical artifact to make it a truly useful and enabling instrument for one's activity (notion of situated acceptance), the individual can be expected to experience difficulties in carrying out work in which he or she recognizes himself or herself and for which he or she is recognized. There is therefore a link between successful technologies and successful work, between a good quality technological environment and a good quality of work, and between the benefit of appropriate technical environments and the well-being of their users.

Introduction written by Marc-Eric BOBILLIER CHAUMON.

New ways of working are also accompanying the deployment of these emerging technologies: dematerialization of work processes, design of dynamic work spaces, automation of intellectual tasks, and the renewal of human–machine collaboration modes with (high-level) tasks that are delegated to systems, distributed among several actors and broken down into multiple spaces/times (nomadism, home office, coworking). ,These new ways of acting also correspond to new ways of suffering, which are expressed by cognitive and psychic loads of a new order and by potential socio-professional and organizational constraints (obsolescence of skills, renewal of professions, etc.).

While technologies can therefore re-enhance work and re-qualify the individual, they can also contribute to distorting the activity and divesting the subject of everything that make sense to them: in their professional practices and ties as well as in their room for maneuver and relationship with work. The dematerialization of activity can therefore be to the detriment of the employee and their work. This is because ICTs are implemented to either replace the individual or appropriate that which represents the core of their activity: that which has meaning and makes sense. However, these tools imply reconfigurations and requirements that destabilize work and weaken the individuals and collectives in place. The question of introducing ICTs and their constant renewal in organizations therefore fundamentally refers to the place and role that these devices play in the activity and its implementation conditions (factors of occupational health and well-being), as well as the way in which human factor specialists (occupational psychologists, sociologists, ergonomists, etc.) can grasp them in their interventions.

Considered as both a current and prospective reflection attempting to grasp the logic and modalities of the digital transformations taking place, this book has several goals:

1) to take stock of the current situation and define precisely what these new technologies and new working methods are that are being deployed in companies;

2) to understand the role they can play in organizational effectiveness and human efficiency;

3) to discuss the role of these technologies in the development (or prevention) of activity and their potential impact on the conditions for exercising the activity, the sources of well-being and the quality of life at work (occupational health);

4) to reflect on the approaches and modalities of support for these digital transformations, based on an anthropocentric, participatory and inclusive approach.

A total of 18 contributions provide epistemological support (from a theoretical and methodological point of view) and empirical illustrations (from field surveys) for these lines of thought. They come from teacher-researchers and practitioners who have been conducting interventions on the deployment of technologies in work organizations and consider the analysis and support of these digital transformations.

These perspectives provide multidisciplinary and complementary insights – *in the field of occupational psychology, ergonomic psychology, ergonomics, sociology of uses, management sciences, etc.* – on how these technologies are transforming work systems and how they can be better managed.

Thus, conceived both as a work of analysis and reflection on the socio-technical changes affecting the world of work in the future and in the making, and also as a handbook for understanding the major technological innovations and the approaches to support digital transformations in progress, each chapter – grouped into four main parts – enriches our understanding of the dynamics at work.

1) The first part provides *an overview of the digital transformations and the various emerging technologies* that are spreading throughout organizations.

Chapter 1 by **Marc-Eric Bobillier Chaumon** provides an overview of the various emerging technologies that are gradually being seen in companies, and discusses their paradoxical uses and impacts on work. Bobillier Chaumon also discusses how he believes the human and social sciences should accompany the design and deployment of these technical innovations so that they support business and promote the well-being of employees. **Chapter 2** by **Nadia Barville-Deromas** and **Marc-Eric Bobillier Chaumon** focuses on the topic of collaborative technologies, and more specifically on the corporate digital social networks that are being massively spread throughout organizations in order to make communication fluid and to optimize synergies between professions and services. The authors question the psychosocio-organizational stakes that such cooperative platforms can represent on collective activity and mobilized skills. **Chapter 3** by **Camille Sagnier**, **Émilie Loup-Escande** and **Gérard Vallèry** proposes a critical analysis of immersive technologies based on virtual and augmented reality. The authors illustrate not only the contributions but also the constraints that these devices impose on the activity and the users, especially in the industrial sector. **Chapter 4** by **Sandrine Berger-Douce** gives an account of the situation of small and medium enterprises (SMEs), as well as their strengths and weaknesses with regard to robotization. It also proposes the means of support for these organizations and their employees, who prove to be poorly prepared or not at all prepared. **Chapter 5** by **Lydia Martin**, **Julian Alvarez** and **Antoine Taly** gives us feedback on the serious game technique for learning professional behaviors, and shows how debriefing and work group devices can promote learning transfers in work situations. Finally, **Chapter 6** by **Tamari Gamkrelidze**, **Moustafa Zouinar** and **Flore Barcellini** defines the scope of action and the possible development of artificial intelligence (AI). The authors question the place and role of AI in professional activities and in the modalities/conditions of interaction between these innovative devices and humans.

2) The following thematic part focuses on *the new ways of working* caused or accelerated by digital transformations, providing analysis and feedback from various research studies.

Thus, **Chapter 7** by **Emilie Vayre** reports on the psychosocial and socio-professional issues of telework deployment within work organizations, based on a summary of various research works and publications. The author also suggests steps that can be taken to mark out the process of deploying telework in work structures. **Chapter 8** by **Elodie Chambonnière** and **Jacqueline Vacherand-Revel** is based on an ethnographic study in the building sector that deals with the reconfiguration of pre-existing management practices of various key players on the building site and occupational risk prevention activity, following the introduction of a business application. The authors demonstrate the advantage of an ethnographic approach that immerses itself in the heart of this activity in order to be able to understand the multiple determinants of the evolution of these newly digitized managerial practices. **Chapter 9** by **Anne-Cécile Lafeuillade**, **Flore Barcellini**, **Willy Buchmann** and **Tahar-Hakim Benchekroun** focuses on collaborative robotics (cobotics) and the way it is considered and determined by the organizational projects of managers of SMEs. The authors demonstrate the importance of starting from the reality of the activity to shape the “desires” of transformation, a reflection similar to that of **Irène Poidi**, **Marc-Eric Bobillier Chaumon** and **Jacqueline Vacherand-Revel** in **Chapter 10**. These authors show how technological artifacts are mobilized at different stages and for different functions in the process of creating innovative companies by young entrepreneurs. These technical devices then take on a status and a particular trajectory of use depending on the situations and activities in which they are employed. Finally, **Chapter 11** by **Maria Ianeva**, **Raluca Ciobanu** and **Chiara Lai** on the new physical territories of digital activity proposes to question the outlines and challenges of space as an object of research and action in digital transformations. Their contribution seeks to qualify the psychosocial and organizational implications of these complex socio-technical devices (flexible spaces, unassigned positions), based on research carried out within the IT subsidiary of a large French group.

3) The third thematic part discusses more specifically *the psychosocial, socio-organizational and occupational impacts of technology diffusion in work activity.*

Chapter 12 by **Maxime Besenval** and **Alexandra Bidet** looks at the beginnings of engagement in digital, creative and collective work. Based on an ethnographic study of the sociology of work and uses in the video game design sector, the authors ask how far we can invest in and develop the opportunities offered by digital transformations – in terms of permanent and collective reconfiguration of the product of work – without threatening the possibility of workers engaging in it. **Chapter 13** by **Florence Cros**, **Marc-Eric Bobillier Chaumon** and **Bruno Cuvillier** raises the issue of the obsolescence of the skills of older workers in the face of the challenges of technological change. This chapter illustrates that older people have major experiential and collective resources that enable them to cope with these transformations, and, above all, to regulate technical failures or work situations not covered by the new systems. **Chapter 14**, the last chapter of Part 3, by **Pauline Crouzat** and **Marc-Eric Bobillier Chaumon** sets out, on the one hand, to specify how the collective activity of an aeronautical engineering department can be reconfigured under the impetus of technology and, on the other hand, to examine the manifestations, resources and limits of these collective practices mediated by technology.

4) The fourth and final thematic part covers *the methods and approaches that can be used to understand, design and support digital transformation projects.*

Thus, **Chapter 15** by **Jean-Marc Robert**, **Antoine Martin**, **Mitra Taraghi**, **Clément Colin**, **Masood Maldar**, **Flavie Bonneviot** and **Eric Brangier** aims to explain the place and role of the prospective ergonomics (PE) approach in the development of future products, services, processes and systems. The authors present two illustrations of innovative projects that apply the approach and methods of this perspective. **Chapter 16** by **Laurent Van Belleghem** describes the interest of the activity simulation method in order to fully explain the new work configurations that these transformations foreshadow and to help design them. Two intervention experiments illustrate these ways of anticipating and co-designing probable future work. **Chapter 17** by **Clotilde Coron** and **Patrick Gilbert** deals with the specificities of change management in the context of digital transformations, because they can be an amplifier, or even a catalyst, of social and professional inequalities. The authors present a model of change management that takes into account the specificities of technological change. **Chapter 18**, the last chapter by **Marc-Eric Bobillier Chaumon**, returns to the concepts and models of technological acceptability, which are crucial in the process of adopting and deploying new technologies. The chapter also discusses their contributions and limitations. Based on the situated acceptance approach, he outlines the methods that, in his view, allow us to explore the acceptance of emerging technologies, based on the analysis of acceptable developments related to digitized activities.

PART 1

Towards an Overview of Digital Transformations…

1

Emerging Technologies and Issues for Activity and Occupational Health

The question of the introduction of emerging technologies and their constant renewal in organizations fundamentally refers to the place and role that these devices play in business, as well as how they can affect activity and health at work. Considered as a prospective reflection attempting to grasp the logic and modalities of the digital transformations underway, the ambition of this chapter is: (1) first, to identify and characterize the emerging technologies that are being deployed in the company; (2) to discuss their impact on the activity and well-being of employees, demonstrating the uses and paradoxical effects that these devices may have on work; and (3) finally, to reflect on the psychosocial function of these devices as potential operators of health and development of the profession.

1.1. Introduction

For several years now, we have been witnessing a major reconfiguration of the conditions in which work is carried out, in their nature (intensification, dematerialization, acceleration), in their modalities (remote working, nomadism, fragmentation, virtual teams) and in their reference points (fragmentation and dispersion of activity, multi-activity), due to the deployment of innovative technical solutions which are multiplying. The impact on the human factor is not neutral. The implementation of these emerging devices can, at best, provoke simple transfers of use and learning; they are iterative or incremental technologies. At worst, they can lead to real disruptions in use; they are then disruptive technologies (Bobillier Chaumon 2017). The latter require a profound rethinking of the user experience,

Chapter written by Marc-Eric BOBILLIER CHAUMON.

requiring new ways of doing, thinking, organizing and collaborating at work. New organizational and socio-cognitive models are then required to work with these devices (de Terssac and Bazet 2007; Ajzen *et al.* 2015), with strong implications for the health and well-being of employees (Bobillier *et al.* 2015). Anastassova (2006) and Loup-Escande and Burkhardt (2019) identify four specific characteristics of emerging technologies: (1) an innovative character, an important technological advance, partially achieved or in the making; (2) unclear and undifferentiated uses; (3) several limitations that slow down its massive application; and (4) a promise of transformation of the economic and social context in which it will be introduced.

The goal of this chapter is to question the place and impact that these emerging technologies can have in the often paradoxical transformations and development of activity. The first step will be to identify and characterize the emerging technologies that are being deployed in companies. Then, we will discuss the impact they may have on professionals and their activity.

1.2. From properties to the uses of emerging technologies

Disruptive digital innovation manifests itself through a variety of technologies that are being deployed in a number of business sectors. Thus, collaborative robots (cobots, exoskeletons), communicating, ambient or ubiquitous technologies (Internet of Things), artificial intelligence (voice assistants, decision support systems), evaluative and predictive data exploitation (Big Data), immersive environments (virtual and/or augmented reality) and new modalities of human–machine interaction (haptics, sensory, cognitive technologies of BIM (brain interface machine) type – brain interface machine) find applications in various fields of our professional and socio-domestic life. They can be found with the factory of the future, the hospital of the future, the smart home, autonomous transport, connected health (HCS/home care service, HS/home support; Martineau and Bobillier Chaumon 2017) or in the services offered by digital work platforms (uberization, robotics; Casilli 2019).

In order to understand the impact of technologies on professional activity and to identify research and societal issues that are addressed to the scientific community, it is necessary beforehand to clarify what they cover in terms of uses and practices, as well as what they bring in terms of resources and constraints.

1) Collaborative robots (also called cobots) are assistants, which remain dependent on the intention, gesture or behavior of humans at work. They support the employee in his/her actions and adjust their interventions to those of the professional. It is no longer simply a robotic substitute or a form of mechanical assistance for particular tasks. Here, robotics becomes symbiotic (Brangier *et al.* 2009), that is, it extends (like an extension) the individual by enabling the increase

of human capacities in terms of strength, speed or precision. In this new context of interaction where these mobile and learning systems evolve, new forms of cooperation and human–robot interfaces (HRIs) are to be imagined and deployed. The exoskeleton constitutes a special class of cobots. It is a device for electrically, pneumatically or hydraulically amplifying the movement of each segment of the body. This kind of external skeleton allows movements, load handling and tool management that the body alone would not be able to perform (Claverie *et al.* 2013). The return of sensation is then immediate, and we witness a certain form of global person–machine consciousness, a hybrid body schema, or what Merle (2012) calls "the illusion of uniqueness". However, we must be careful, because these exoskeletons that are grafted onto the human body strongly constrain body movements and gestures. It is more about adjusting to this "mechanical corset" than the other way around: the actions of the body can be repressed or even prevented, with possible impacts on the physical health of employees (increase in musculoskeletal disorders – MSDs).

2) Ambient, ubiquitous or pervasive technologies constitute a second class of these emerging innovations, more generally known as the "Internet of Things" (IoT) or connected/communicating objects. These discrete (because they are non-intrusive) technologies are integrated into everyday objects (Nehmer *et al.* 2006). They seek to capture activity and trigger the appropriate actions, without the need for human intervention (Gossardt 2017) (e.g. the thermostat that communicates with the personal calendar to turn on the heating at the appropriate time). These can be sensors scattered throughout the living or working space to record physical activities (quantified-self; Zouinar 2019), to inform the individual and thus fight against sedentariness: people are then made aware of what they are doing, what they are not doing – well or not well enough – and what they should do better (concept of "technological persuasion", nudge technology or captology; Fogg 2002). There are also digital tracers integrated into production lines to evaluate, in real time, the conformity of professional actions to expected standards. Maintenance is no longer merely corrective, it becomes predictive. That is, we are able to react before the error is made by measuring the nut and bolt too, if it is wrong or not tight enough. This is the idea of the connected factory of the future (Factory 4.0).

3) Artificial intelligence (AI) is highly sophisticated algorithmic programs based on artificial neural networks. They aim to solve specific problems for which human beings use their cognitive abilities (Zouinar 2020). What constitutes the strength of these devices is their "reasoning" power, which is based on deep learning or machine learning (see Chapter 6). These particular machine learning techniques make it possible to analyze, extract and classify large amounts of data. These programs are used to make diagnoses (skin cancer detection is more reliable than human expertise), make decisions (90% of stock market orders are now made by AI systems, such as high-frequency trading) or "naturally" assist humans in their daily

activities, for example, Google's voice assistant, capable of making appointments by phone, voice-recognition systems (connected speakerphone) or chat-bot systems that automatically answer customers' questions on websites. These devices are "artificial intelligence" because they are capable of some form of learning, or even "intuition" to deal with complex and sometimes unprogrammed situations. This is the case of the autonomous car, which has to manage a wide range of unforeseen events and incidents on its course. However, it should also be recalled that although AI is capable of learning, evolving and taking initiatives[1], it is incapable of giving meaning to what it sees or does or to the information it processes. Concretely, it can recognize the presence of a cat in millions of photos, but it does not know what a cat is. Symbolic processing and cognitive flexibility remain for the moment the prerogative of humans.

4) In order to function, AI must be able to access large volumes of data provided by, among others, Big Data. Big Data refers to the ability to produce or collect digital data, store, analyze and visualize them (Cardon 2015). The traces we leave on the Internet ("likes", comments), the data sent by connected objects... on all aspects and at all times of our lives constitute a considerable mass of information[2] on how we act, think and even on what we experience and feel (Dagiral *et al.* 2019). However, this raw data have little meaning and value as it stands. The challenge is to assign value to them through correlation, in order to transform them into information, and then into knowledge about the subject, that is relevant, useful and focused data. These are "smart data". Combined with predictive models, these systems are capable of evaluating and anticipating individual behavior in some detail, or even seek to modify attitudes and decisions (as in the case of the Cambridge Analytica scandal[3]).

5) Immersive environments (based on virtual, augmented and tangible reality) consist of immersing a person in an artificial environment, by stimulating their sensory (via sound, vision, smell), cognitive (information and decision-making) and sensory-motor (haptics, gestures) modalities via appropriate digital devices (3D

---

1 In companies, these systems are used (1) to relieve employees of repetitive cognitive tasks (e.g. in a bank branch, IBM's Watson system automatically answers customers' emails to relieve the customer advisors' relational work) or (2) to support or even replace them in complex tasks. Examples include navigational aid systems (automatic pilot in aircraft) or expert systems in the banking field (to evaluate the best financial investments) or in the legal field (for example, finding the best case law by compiling thousands of applied legal decisions).

2 Today, more than 98% of the information produced by our companies is stored in digital form, whereas in 1986 only 1% of data was in this format.

3 The name of the British company that has been accused of influencing the vote in the 2016 US elections through the improper use of personal data from tens of millions of Facebook accounts.

headsets, haptic gloves, etc.) (see Chapter 3). This world can be imaginary, symbolic or simulate certain aspects of the real world. Different types of immersive environments can be distinguished:

– Virtual reality makes it possible to extract ourselves from physical reality in order to virtually change time, place and/or types of interaction. It gives a person the opportunity to engage in sensory-motor and cognitive activity in an artificial, digitally recreated world (Fuchs *et al.* 2006). It allows us to simulate what we would do in a normal and real situation. These devices are often used in the field of vocational training: employees find themselves in situations close to their actual working conditions, which are difficult to reproduce (Ganier *et al.* 2013). For example, it may be necessary to simulate the altitude to carry out a technical intervention at the top of an electrical pylon, or to carry out a difficult operation on a patient, with medical complications. Virtual reality can also be used in the medical field to treat anxiety-provoking situations (treatment of post-traumatic stress disorder (Moraes *et al.* 2016) or delusions of persecution (Freeman *et al.* 2016)).

– Augmented reality consists of adding/enriching virtual information above the real physical environment using a video headset, a computer or any other projection system (Marsot *et al.* 2009). There are applications for productive or maintenance tasks (such as indicating to the operator very precisely the location of rivets to be screwed on the aircraft's cabin: the targets are then projected virtually on the surface) or when the operator wearing a headset is presented virtually with the sequences and location of the various operations to be performed to change a part on a large industrial machine (the steps and circuit to be changed appear superimposed on the engine). We therefore interact with the virtual in order to know how to act on the real thing.

– Augmented virtuality (tangible environment) consists of integrating real entities into a virtual environment; both can interact together (Fleck and Audran 2016). For example, an architect will physically manipulate models of houses in a virtually recreated living space in order to evaluate the best exposure to sun and wind and thus calculate possible energy losses. In this case, we interact with reality to act on the virtual.

The integration of these various immersive technologies (virtual reality, augmented reality, augmented virtuality) is called mixed reality (Moser *et al.* 2019).

All these new generations of technologies are thus intended to replace/improve all or part of human functions (physical, sensory and/or cognitive). The objective is to optimize the capacities at work (learning, understanding, decision-making, action, etc., both individual and collective) and to make work processes more efficient and effective in order to increase reactivity and profitability. According to a very deterministic approach, it also appears that the choice of such systems aims at the emergence of a working model oriented towards individual excellence, organizational agility, collective intelligence and also an efficient mutualization of the activity (between humans and machines). This would also explain the enthusiasm of companies for such systems, as Champeaux and Bret (2000, p. 45) have already mentioned, for technologies that are now more traditional: "Adopting them is no longer an opportunity, but an obligation. It is no longer a question of whether we are going to go there, but of how we are going to go there, that is, with what strategy, what investments, what objectives".

However, while it appears that technology can affect certain dimensions of the activity, it cannot determine or shape it according to predefined and expected patterns. There is no technological determinism in the strict sense of the term. In other words, a technological innovation does not in itself impose a single type of organization or business model, but makes various forms of it possible. It is indeed the use (i.e. the conditions of use of the tool – individual, collective, organizational, etc., the project and the experiences of the user…) and not the intrinsic characteristics of the technology that will determine its effects, which can therefore be contrasted. It is these paradoxes that we will now examine in the following section.

1.3. Five paradoxes of the diffusion of technologies in/on the activity

1.3.1. *Sense of loss of control over the activity vs increased control over the activity*

The multiplication of digital media at work (shared digital schedules and spaces, messaging and social networks, collaborative platforms, reporting tools) is accompanied by a requirement for permanent availability and reactivity. The activity is thus increasingly determined, punctuated and emphasized by the various digital injunctions, warnings and solicitations. In the research we conducted on the impact of technology on the management profession (Bobillier Chaumon *et al.* 2018), we identified a category that we called "self-service or dispossessed managers". They felt deprived of their ability to define or control their work schedule (work "imposed" by shared schedules) or simply to keep and achieve the objectives they had set for themselves during the day (work "prevented" by the many digital interruptions that required them to complete the current activity in order to initiate a new, unplanned task). This feeling of loss of control over the activity was paradoxically consubstantial with the increase in control that employees experienced in their activity.

Indeed, more and more so-called prescriptive (Bobillier Chaumon 2017) or info-normative (Frenkel *et al.* 1992) technical systems determine and frame the work to be done, but are also capable of assessing whether the work has been done well, that is, whether it is compatible with labor standards and norms. The software packages that manage the dialogue between advisors and customers in call centers, the voice-picking devices that dictate each action to the warehouse agent to make a pallet, thus establish what must be done, how it must be done, according to what timeframe and with whom.

1.3.2. *Invisibility vs visibility of the activity*

In an activity that tends to become dematerialized, less tangible and less comprehensible, more distributed between human–machine systems, the individual finds it increasingly difficult to assess what his or her actual contribution to work is. He/she thus intervenes at a given moment or on a particular sequence of the work process, without being able to judge what he/she really brings, his/her added value in the task, or the various impacts of his/her professional conduct: in terms of customer satisfaction, improvement of the work process or efficiency of the actions undertaken.

This lack of visibility on the activity gives the feeling of a "truncated" action, that is, the professional has difficulty not only in evaluating who he/she serves (is it a customer, a robot, an automated procedure?), but also what he/she is doing the work for and especially what his/her work is used for. This sense of usefulness is one of the fundamental attributes of professional recognition (Dejours 1993). It is based on self-actualization in the social field of valorization: of oneself by oneself, and of oneself by others: that is, by the way I recognize myself in what I do and also in the way others perceive me in conjunction with what I do. Professional identity also asserts itself as the framework of psychological health. Not being able to identify with what we do and to be recognized (by our peers, those in charge) in what we do, weakens our identity (the way we consider, appreciate and value ourselves) and consequently affects our well-being.

Conversely, these same systems can contribute to increasing the visibility of work through systems and indicators (reporting, indicators, Big Data, connected objects) that make it possible to (re)trace and evaluate in almost real time all the work that the individual does, and above all the work that he or she does not do, in relation to the standards performed. With the connected factory, for example, we will have powerful means that recount who does what, when and within what timeframe. This panoptic world (Bain and Taylor 2000), constantly supervised by these technologies, also makes the individual more transparent and measurable and therefore more predictable and controllable as well. In other words, one has the impression that these ICTs give one more control and more capacity for action over one's human behavior (through better self-knowledge). But we are also creating the conditions for voluntary servitude with regard to these devices: which tell us how to act, at what pace to live, and what the right attitude to have is.

1.3.3. *Increase in virtual teams vs isolation of employees*

A third contradiction concerns the increase in virtual teams/collaborations, which can paradoxically be accompanied by the isolation of employees. This also reflects the opposition between "*collective work*" and "*a work collective*" (or teamwork) (Caroly and Clot 2004).

Collaborative work platforms and corporate digital social networks are thus being deployed, which ultimately aim more at prescribing collaborations between employees than at supporting existing cooperations. Without this bias, the company seeks to optimize collective intelligence, transversality and decompartmentalization between departments and teams, which are the new spearheads of matrix and agile organizations (Barville *et al.* 2018; see Chapter 2 in this book).

The aim of these systems is to order or even impose special forms of cooperation between employees in order to increase exchanges and collaborative work remotely (Crouzat and Bobillier Chaumon 2017; see Chapter 14 in this book). However, by seeking to organize collective work – at the expense of collective work – these technical environments create artificial collectives (Crapeau 2001) composed of individuals opportunely linked according to the needs of the organization and projects. However, this does not in any way promote the feeling of belonging to a professional community, to a work collective, or teamwork.

It should be remembered that this working group is the indispensable framework for regulating work constraints. The collective refers to a community of actions and thoughts, essential in the relationship between the individual and the work situation (Lhuilier 2006). It creates a sense of belonging and identification with a professional community. The role of the collective is therefore essential, because it is both a resource and a support (psychological, cognitive, social, professional) for individual action and a "social shock absorber" for organizational pressures. The absence or decline of these groups is partly responsible for the suffering of employees who are then unable to manage the various risks of the activity and to cope with the demands of the organization.

With these collaborative environments, if people feel like they are working virtually together, they do so by being more alone in front of their computers. This contributes to creating what Clot (2010) calls a collection of individuals, that is, an arrangement of interchangeable skills, linked together by fake relationships, and where individuals are exposed to isolation: they do not know who to rely on, who to turn to in case of a problem, who to open up to in case of a hard blow. This isolation can be compounded by a feeling of loneliness when one feels excluded or even banished from the working group to which one belongs. One feels alone among others in a way. Marc and Ladreyt (2019) speak of distressing loneliness to describe "a subjective experience suffering from delirium associated with a perceived lack of social support. The subject faces a solitary and constrained face-to-face encounter with himself/herself. This form of loneliness is pathogenic. It is an experience conducive to restlessness" (p. 262), which is accompanied by a withdrawal into oneself, a voluntary abandonment of all ties with others. The result is a double exclusion "of oneself among others and of others in oneself".

This distressing loneliness can appear when the organization "corrupts" the use of certain technological artifacts by putting employees in competition with each other, which can be harmful to existing collectives. "Surviving together" then gives way to "every man for himself". Indeed, as we have seen previously, certain tools seek to trace the activity and to report on the task performed using various numerical indicators. However, these indicators (often partial because they are based on the performance performed) prove to be inappropriate and insufficient to testify to the specificities and requirements of the real activity: namely, all the trials, errors, bewilderment, etc. necessary to complete the work. Management will, however, use these data to encourage emulation between employees. However, these individualized evaluations push individuals to judge, compare, confront and finally come head to head with each other between competitors (and no longer colleagues) at the risk of jeopardizing the relationships of mutual aid and trust, which are necessary for social cohesion in the activity. The other is then seen as an adversary that I must protect myself from and keep at a distance. This competition with others can also be doubled with competition with oneself when the individual seeks to maintain a level of excellence in the activity (de Gaulejeac and Hanique 2018), which in the end turns out to be very deleterious, even deadly.

1.3.4. *Nomadism vs a sedentary lifestyle at work*

The office is no longer the major or exclusive determinant of work. Thanks to digital environments (permanent office, "office in the pocket"), it has been exported, largely outside the walls of the company (work without borders) and has freed itself from the temporal, spatial and organizational barriers of the company. While some employees carry out an activity from their home (home office, teleworking; Vayre

2019), others tend to engage in a nomadic type of activity by alternating between workplaces (customer base, hotel, transport, third party places such as coworking areas, etc.) and different schedules (part of the work being carried out in "concealed" time, i.e. on transport or during other waiting times) (Vacherand-Revel 2017; Lutumba *et al.* 2020). These employees thus find themselves more or less voluntarily immersed in environments that each have their own requirements and that demand particular practices and uses. Thus, depending on the digital demands, employees must improvise their professional activities in situations that they do not always control, and which do not necessarily lend themselves to the task to be carried out with the technical system (Bonneau and Enel 2018). The different roles and work contexts then interpenetrate each other and this can create conflict (Marks and MacDermid 1996). Concretely, the individual may be psychologically and/or behaviorally engaged in one area/situation, but physically and temporally present in another area. Being able to carry out an activity in an inappropriate and potentially disruptive environment requires increased attention and focus (Popma 2013).

While nomadism is one of the characteristics of modern work, which technology has undoubtedly contributed to and has accentuated and accelerated it, sedentary life is paradoxically another facet of these digitized situations.

Indeed, there are still a number of economic sectors in which employees remain glued to their computers to carry out almost all the tasks inherent in their professional activity: from mediated interactions between colleagues (remote offices, being away from the workplace) to remote meetings, from information management (research, archiving) to training (e-learning) and even for moments of relaxation and social breaks (surfing on the web and social networks) (Debrosses 2019). It should be noted that this sedentary lifestyle has a major health cost since, according to epidemiological studies, it causes various major organic and physical disorders and causes mortality greater than that caused by cigarette consumption (Wen and Wu 2012). A study by the American Cancer Society indicated that the mortality rate of a person sitting more than six hours a day is 20% higher than that of a person sitting only three hours a day (Patel *et al.* 2010).

1.3.5. *Detachment from activity vs proximity of work*

A final contradiction is the tension between distancing ourselves from work, on the one hand, and being very/too close or even experiencing a lack of privacy, on the other hand.

Indeed, in parallel with the multiplication of interfaces between a human and their work, we are witnessing a distancing between humans and the object of their activity (Dodier 1995). The individual no longer acts directly on the product of their work (object to be transformed) or interacts with the person to be supported (client to be satisfied), but goes through a technical mediator (e-mail, collaborative or robotic tool) which will act on their behalf. However, this technical artifact requires other ways of representing ourselves, organizing and implementing our work (Norman 1994; Rabardel 1995). It is also a dematerialized work that becomes more abstract, more symbolic, less tangible, because it is represented by curves, signs, values, codifications, etc. The professional must then base himself/herself on these new indices of the activity to enable him/her to understand the process in progress and to adjust his/her interventions accordingly.

In this activity mediated by technologies, the reference points are no longer sensory (linked to noise, to a characteristic smell in the activity), and no longer rest on interactional and/or physical/haptic material bases (such as irregularity spotted in a work process, a distrustful attitude identified during customer negotiations). These indicators are now signified by matrices, which the professional must interpret to give them a meaning, a value (Baril 1999). He/she thus loses the intimate knowledge of the product of the activity: its specificities, its properties, its reactions, etc. Consequently, this loss of meaning – through media detachment – is accompanied by a loss of meaning (Baudin 2017), that is, one no longer recognizes oneself in the product/service being made.

Another consequence of this remoteness concerns the maintenance of competence and of the professional gesture which, when they are no longer maintained, withers, weakens and deteriorates. There is thus a risk of losing the acuity of analysis (based on information taken from the field), the dexterity and assurance of conduct, and the skill and finesse of movement. Beyond the decline of the gesture, the confidence itself in the gesture can also be dulled. This refers to Bendura's (2007) sense of personal efficacy (SPE), which is reflected in the fact that one no longer feels capable and confident to be able to ensure the required professional conduct. We can give the example of those surgeons who, by dint of using mechanized extensions (interfaces and robotic arms), no longer have the same knowledge of the body and the patient's reactions (through sensory and physical cues) and no longer develop the same motor skills that require the manipulation of the scalpel (Wannenmacher 2019). It is other skills and abilities (perceptive, motor, collective) that are mobilized to use alongside the technological system (Seppänen *et al.* 2017).

This physical and sensory detachment from the object of one's activity is paradoxically accompanied by greater promiscuity with work. Indeed, work has never been so present, pressing and oppressive through the use of technology. Clients, colleagues, managers and work procedures are becoming omnipresent and overflow from the professional sphere to invade domestic, personal and third party areas (El-Wafi *et al.* 2017). Messages, digital solicitations and warnings arise at any time and in any place, promoting intrusion, the hybridization of personal and professional spaces and fostering the feeling of excess and over-connection (Morand *et al.* 2018). This feeling is also known as telepressure (Dose *et al.* 2019).

1.4. Conclusion

Emerging technologies, by promising new and almost infinite computational capacities, reasoning possibilities and modes of action, induce an all-powerful imagination: they are then perceived or presented as omnipotent, because they are omniscient (reasoning power of AI) and omnipresent (supervision of activities by smart data and connected objects). They formulate an almost superhuman injunction to be able to control them. And the individual, a mere mortal, can only be, at best, the discreet substitute or the passive auxiliary of these systems, and at worst, they'll be totally eradicated from the socio-technical equation that is unfolding.

What is also particular with these emerging technologies is to consider the omnipotence of these tools as determining and structuring the organization's project: factory/hospital of the future, collaborative work platform, smart home, teleworking, collective intelligence, organizational agility, work tuberization, etc. We are in what could be called a "technological absolutism" or an essentialist vision of technology: first and foremost, one considers what it "is" and "must do", instead of what it is capable of "doing" or "undoing", "bringing or removing", that is, its value, the meaning it takes on in the activity. It is the only one that holds the truth, the knowledge, the expertise about the work. Seen as performative systems, these tools are the Alpha and Omega of organizational efficiency and individual and collective innovation.

However, despite the innovations and sophistication that characterize these technologies, it should be remembered that the vast majority of these devices remain disconnected from the socio-professional realities that employees experience. They are more oriented towards the search for technological performance, profitability and socio-economic efficiency, which disqualify from the outset any subjective commitment and deny any human initiative that could be detrimental to the organizational project. For companies, these human practices thwart their action plans and/or threaten the quality standards promoted by technological environments. In this system of activity-mediated human–machine, it is the individual who is the variable of adjustment and it is up to him/her to find an acceptable compromise of functioning, with the rules of action imposed by the device. Moreover, as we have also seen, the deployment of such environments is also accompanied by tensions and contradictions in the very way of living and acting with these tools, of building our practice and developing our profession.

Ultimately, in this human–machine collaboration to be constructed, it is a question of moving from "incantatory technologies", which aim to provoke quasi-automatic, even magical effects, by their mere presence; to "embodied" technologies which have meaning for employees and which give meaning to the activity that is being done. They aim not only to satisfy the needs of professionals, but also to support their initiatives, sustain their activity and encourage the development of their skills and autonomy. It is under these conditions that these technologies will become acceptable (Bobillier Chaumon 2016, and see Chapter 18 in this work) by being instruments at the service of professions and acting as occupational health operators.

1.5. References

Ajzen, M., Donis, C., and Taskin, L. (2015). Kaléidoscope des nouvelles formes d'organisation du travail : l'instrumentalisation stupide d'un idéal collaboratif et démocratique. *Gestion 2000*, 32(3), 125–147.

Anastassova, M. (2006). L'analyse ergonomique des besoins en amont de la conception de technologiesémergente : le cas de la réalité augmentée pour la formation à la maintenance automobile. Paper, Université René Descartes – Paris V.

Andonova, Y. and Vacher, B. (2013). Nouvelles formes de visibilité des individus en entreprise : technologie et temporalité. Communication et organisation, 44(2013), 5–14.

Bain, P. and Taylor, P. (2000). Entrapped by the electronic panopticon? Worker resistance in the call centre. *New Technology, Work and Employment*, 15(1), 2–18.

Bandura, A. (2007). *Auto-efficacité : le sentiment d'efficacité personnelle.* De Boeck, Brussels.

Baril, R. (1999). Les transformations du travail des opérateurs de raffinerie de pétrole : le passage des cadrans aux écrans. *PISTES*, 1(1) [Online]. Available: http://journals.openedition.org/pistes/3843 [accessed 27 April 2020].

Barville-Deromas, N., Bobillier Chaumon M.E., Cippelletti, E., Troyano, V., and Aguercif, S. (2018). Quand les technologies s'incarnent dans l'activité : le cas des réseaux sociaux numériques d'entreprise. *Psychologie du travail et des organisation*, 24(4), 325–339.

Baudin, C. (2017). La perte des sens et du sens des corps technicisés. *Actes FISO (Festival International de SOciologie), la fabrication des corps au 21e siècle*, (2), 1–15, Epinal, France.

Bobillier Chaumon, M.E. (2016). Acceptation située des TIC dans et par l'activité : premiers étayages pour une clinique de l'usage. *Psychologie du travail et des organisations*, 22(1), 4–21.

Bobillier Chaumon, M.E. (2017). Du rôle des TIC dans la transformation digitale de l'activité et sur la santé au travail. *La Revue des conditions de travail*, ANACT, 6, 16–24.

Bobillier Chaumon, M.E., Brangier, E., and Fadier E. (2015). Usage des technologies de l'information et bien-être au travail. In *Pathologie professionnelle et de l'environnement.* Fadier, E. (ed.). Editions Elsevier Masson, Paris.

Bobillier Chaumon, M.E., Cuvillier, B., Sarnin P., and Vacherand-Revel, J. (2018). Usage des TIC et évolutions des pratiques socioprofessionnelles des cadres : quels repères pour le métier et quelles incidences sur la santé ? *Pratiques psychologiques*, 34(4), 349–373.

Bonneau, C. and Enel, L. (2018). Caractériser le métatravail des nomades numériques : un préalable à l'identification des compétences requises. *Lien social et politiques*, 81, 138–155.

Brangier, É., Dufresne, A., and Hammes-Adelé, S. (2009). Approche symbiotique de la relation humain-technologie : perspectives pour l'ergonomie informatique. *Le travail humain*, 72(4), 333–353.

Cardon, D. (2015). *A quoi rêvent les algorithmes : nos vies à l'heure des Big Data.* Le Seuil, Paris.

Caroly, S. and Clot, Y. (2004). Du travail collectif au collectif de travail : développer des stratégies d'expérience. *Formation Emploi*, 88, 43–55.

Casilli, A. (2019). *En attendant les robots : enquête sur le travail du clic.* Le Seuil, Paris.

Champeaux, J. and Bret, C. (2000). *La cyber entreprise*. Dunod, Paris.

Claverie, B., Le Blanc, B., and Fouillat, P. (2013). La cobotique, la robotique soumise. *Communication et organisation*, 44, 203–213.

Clot, Y. (2010). *Le travail à coeur. Pour en finir avec les risques psychosociaux.* La Découverte, Paris.

Craipeau, S. (2001). *L'Entreprise commutante, travailler ensemble séparément.* Hermès, Paris.

Crouzat, P. and Bobillier Chaumon, M. (2017). Technologies collaboratives : usages et impacts au sein de l'activité collective. *La Revue des conditions de travail*, ANACT, 6, 145–152.

Dagiral, E., Licoppe, C., Martin, O., and Pharabod, A.S. (2019). Le quantified self en question(s) : un état des lieux des travaux de sciences sociales consacrés à l'automesure des individus. *Réseaux*, 216(17), 21–54.

De Gaulejac, V. and Hanique, F. (2015). *Le capitalisme paradoxant : un système qui rend fou*. Le Seuil, Paris.

Debrosses, K. (2018). Le comportement sédentaire au travail : de quoi parle-t-on ? *Hygiène et sécurité du travail*, INRS, 252, 6–10.

Dejours, C. (1993). Intelligence pratique et sagesse pratique : deux dimensions méconnues du travail réel. *Education permanente*, 116, 47–70.

Dodier, N. (1995). *Les hommes et les machines : la conscience collective dans les sociétés technicisées*. Editions Métailié, Paris.

Dose, É., Desrumaux, P., and Rekik, M. (2019). Télépression au travail, relation entre l'utilisation de la messagerie électronique au travail et les échanges leader-membres : rôles de la reconnaissance et de la charge de travail. *Le travail humain*, 82(2), 151–181.

El Wafi, W., Brangier, E., and Zaddem F. (2016). Usage des technologies numériques et modèles de la perméabilité des frontières entre la vie personnelle et la vie professionnelle. *Psychologie du travail et des organisations*, 22(1), 74–87.

Fleck, S. and Audran, J. (2016). Réalité augmentée et interfaces tangibles, quels liens avec la formation par simulation ? *Recherche et formation*, 82, 85–90.

Fogg, B.J. (2002). Persuasive technology: Using computers to change what we think and do. *Ubiquity*, 5(1), 89–120.

Freeman, D., Bradley, J., Antley, A., Bourke, E., De Weever, N., Evans, N., and Clark, D.M. (2016). Virtual reality in the treatment of persecutory delusions: Randomised controlled experimental study testing how to reduce delusional conviction. *The British Journal of Psychiatry*, 209(1), 62–67.

Frenkel, S., Korczynski, M., Shire, K., and Tam, M. (1998). Beyond bureaucracy? Work organisation in call centres. *The International Journal of Human Resource Management*, 9(6), 957–979.

Fuchs, P., Berthoz, A., and Vercher, J.-L. (2006). Introduction à la réalité virtuelle. In *Le Traité de la réalité virtuelle*, Fuchs, P., Berthoz, A., and Vercher J.-L. (eds). Les Presses de l'Ecole des Mines de Paris, Paris.

Ganier, F., Hoareau, C., and Devillers, F. (2013). Évaluation des performances et de la charge de travail induits par l'apprentissage de procédures de maintenance en environnement virtuel. *Le travail humain*, 76(4), 335–353.

Gossart, C. (2017). Les enjeux sociétaux des objets connectés. *Terminal*, 120 [Online]. Available: http://journals.openedition.org/terminal/1500 [accessed 4 May 2020].

Lhuilier, D. (2006). *Cliniques du travail*. Érès, Toulouse.

Loup-Escande, E. and Burkhardt, J.M. (2019). Évaluer l'utilité dans le contexte des technologies émergentes pour identifier des besoins latents : éléments issus d'une analyse des interactions en situation d'usage. *Activités*, 16(2) [Online]. Available: http://journals.openedition.org/activites/4554 [accessed 4 May 2020]. DOI: https://doi.org/10.4000/activites.4554.

Lutumba, P., Bobillier Chaumon M.E., and Miossec, Y. (2020). Quand l'activité s'affranchit des frontières spatiales et temporelles : le cas des éducateurs spécialisés nomades. *Actes du 55ème Congrès de la SELF, l'activité et ses frontières. Penser et agir sur les transformations de nos sociétés*. Paris, 16–18 September 2020.

Marc, J. and Ladreyt, S. (2019). Isolement au travail et solitude professionnelle. In *Psychologie du travail et des organisations : 110 notions clés*, Valléry, G., Bobillier Chaumon, M.E., Brangier, E., and Dubois, M. (eds). Dunod, Paris.

Marks, S.R. and MacDermid, S.M. (1996). Multiple roles and the self: A theory of role balance. *Journal of Marriage and the Family*, 58(2), 417–432.

Marle, S. (2012). *Super héros et philo*, Breal, Paris.

Marsot J., Gardeux F., and Govaere V. (2009). Réalité augmentée et prévention des risques. Apports et limites. *Hygienne et sécurité du travail (INRS)*, 214, 15–23.

Martineau, M.L. and Bobillier Chaumon, M.E. (2017). Présence de robots dans les Etablissements d'Hébergement pour Personnes Agées Dépendantes (EHPAD) : mieux vivre la séparation familiale. *Dialogue : recherche sur le couple et la famille*, 217(3), 45–56.

Moraes, T.M., de Andrade, A.F., and Paiva, W.S. (2016). Virtual reality for the treatment of posttraumatic disorders. *Neuropsychiatric Disease and Treatment*, 12, 785–786.

Morand, O., Cahour, B., Bobillier Chaumon, M.E., and Grosjean, V. (2019). Overload of technological connections for communicating at work. In *Proceedings of the 20th Congress of the IEA 2018*. Bagnara S., Tartaglia R., Albolino S., Alexander T., and Fujita Y. (eds). Springer, Cham.

Moser, T., Hohlagschwandtner, M., Kormann-Hainzl, G., Pölzlbauer, S., and Wolfartsberger, J. (2019). Mixed reality applications in industry: Challenges and research areas. In *SWQD 2019: Software Quality: The Complexity and Challenges of Software Engineering and Software Quality in the Cloud*, Winkler, D., Biffl, S., and Bergsmann, J. (eds). Springer, Cham, Switzerland.

Norman, D. (1994). Les artefacts cognitifs. *Raisons pratiques*, 4, 15–34.

Patel, A.V., Bernstein, L., Deka, A., Feigelson, H.S., Campbell, P.T., Gapstur, S.M., Colditz, G.A., and Thun, M.J. (2010). Leisure time spent sitting in relation to total mortality in a prospective cohort of US adults. *American Journal of Epidemiology*, 172(4), 419–429.

Popma, J. (2013). *The Janus Face of the 'New Ways of Work': Rise, Risks and Regulation of Nomadic Work*. ETUI, Brussels.

Rabardel, P. (1995). *Les hommes et les technologies*. Armand Colin, Paris.

Seppänen, L., Kloetzer, L., and Riikonen, J. (2017). Encourager la réflexion en chirurgie robotique : l'autoconfrontation croisée comme instrument de développement du travail interprétatif dans un environnement technologique complexe. *Psychologie du travail et des organisations*, 23(2), 117–136.

de Terssac, G. and Bazet, I. (2007). Les TIC-TAC de la rationalisation : un travail d'organisation. In *La rationalisation dans les entreprises par les technologies coopératives*, de Terssac, G., Bazet I., and Rapp L.L. (eds). Octarès Editions, Paris.

Vacherand-Revel, J. (2017). Le travail en équipe à l'ère de sa médiatisation numérique : figures actuelles de l'innovation organisationnelle. *Psychologie du travail et des organisations*, 23(2), 85–88.

Vayre, É. (2019). Les incidences du télétravail sur le travailleur dans les domaines professionnel, familial et social. *Le travail humain*, 82(1), 1–39.

Wannenmacher, D. (2019). Impacts et enjeux du robot chirurgical dans les blocs opératoires : passer des compétences techniques individuelles aux compétences non-techniques collectives. *Journal de gestion et d'économie de la santé*, 4(4), 316–334.

Wen, C.P. and Wu, X. (2012). Stressing harms of physical inactivity to promote exercise. *The Lancet*, 380(9838), 192–193.

Zouinar, M. (2019). Théories et principes de conception des systèmes d'automesure numériques : de la quantification à la régulation distribuée de soi. *Réseaux*, 216(4), 83–117.

Zouinar, M. (2020). Évolutions de l'Intelligence Artificielle : quels enjeux pour l'activité humaine et la relation humain-machine au travail ? *Activités*, 17(1) [Online]. Available: http://journals.openedition.org/activites/4941 [accessed 26 April 2020].

2

Collaborative Work Platforms: Challenges for Business Development

This chapter is based on research that addresses the psychosocial–organizational issues that can arise from the use of a digital enterprise social network (DESN). After defining the hybrid modalities that characterize contemporary work, we will demonstrate that the activity of networking reconfigures the engagement and activity structuring in asynchronous and distributed collaborations. We will rely on the concept of shared collaborative intentionality to qualify these new modalities of engagement and to characterize the new collaborative skills developed by professionals in order to increase their power to act.

2.1. Introduction

Collaborative platforms and digital enterprise social networks (DESNs) bring together a set of technologies (ICTs) (wikis, forums, instant messaging, micro-blogs, document sharing and content co-design spaces). They are similar to the digital social networks used privately (Facebook, Twitter, LinkedIn, etc.), and constitute new types of technologies. These are dedicated to communication and collaboration between companies through the development of new communication modalities, the creation of remote content, the sharing of opinions and the capitalization of information. Two objectives are targeted: (1) to optimize interactions; and (2) to facilitate collaborations between actors who are physically distant or who cannot collaborate at the same time (Zacklad 2006; Barville-Deromas 2014).

Chapter written by Nadia BARVILLE-DEROMAS and Marc-Eric BOBILLIER CHAUMON.

In this chapter, we will continue to use the term digital enterprise social network (DESN). As a technical–organizational device, the DESN aims to virtualize exchanges and sharing while supporting asynchronous collaborative co-writing mechanisms (Zacklad 2015).

2.2. Two organizational challenges: empowering digital transformations and changing work practices

DESNs began to appear in companies in 2008, in a context of recurring organizational change and digital transformations (Cohendet *et al.* 2003; Teulier and Lorino 2005; Engelstätter and Sarbu 2013). Their implementation has accelerated since 2015. They are used in so-called hybrid work organizations, that is, those that combine pyramidal and transversal operations (Engeström 2008a). This hybrid aspect of work organization aims to develop the network effect (a development of efficiency through the decentralization of information control centers). It is a new model of success that combines economic performance and the knowledge economy. In fact, the organization of daily work is based on a multi-centered (several activities in parallel) and reticular (networked, without hierarchical links) structure. Two major changes in the organization of daily work are emerging: the co-configuration of work and the mobilization of social capital.

2.2.1. *The co-configuration of work*

The co-configuration of work consists of the growing need to make very different actors collaborate, those that are both internal and external to the company. The aim is to ensure the production of a service or product that is tailored to the needs and desires of customers (Engeström 2008b). First applied to the economic context, the co-configuration of work has gradually integrated into companies, instilling new collaborative dynamics that are aimed at creating individual and collective values.

From a work organization point of view, co-configuration requires a great ability to collaborate in a highly variable environment, to adapt, to make it look good (customize) and to be transparent. Professionals must therefore juggle different work purposes (work objects) on a daily basis: responding to the requests of superiors, achieving the objectives set as a team and, at the same time, getting involved in cross-functional work groups with colleagues who do not belong to the same team and with purchasers who are not hierarchical superiors (Edwards 2011).

These mechanisms are supposed to facilitate the articulation of professionals among themselves, in order to better manage meetings and the structuring, capitalization and dissemination of knowledge, based on the information they hold in a professional or personal capacity. Thus, the functionalities of DESNs offer to these professionals dispersed within the company (*belonging to different teams or projects and professions*), the ability to jointly write documents in an asynchronous way, using different formats (*wiki pages, publishing of documents, posting comments*), to express a point of view or an opinion using interactional functionalities (*like, get in touch, subscribe to a group*), and to interact with a group of actors dispersed within the company (*belonging to different teams or projects and professions*), to work together on common subjects.

2.2.2. *Mobilizing social capital to make sharing more flexible*

Social capital is defined as the set of relationships that can be mobilized by an individual or a group and can be considered to be a resource that is comparable to economic capital. It is based on a principle of solidarity that makes it possible to give information to colleagues. It is based on cooperation dynamics that exist outside of formal procedures (Bourdieu 1980; Alter 2009).

In these cross-functional, asynchronous and multi-professional work contexts, the manager's role as an arbitrator tends to disappear. This translates into a net increase in the decision-making autonomy of the employee, who must then develop new, less cognitive and more relational skills (*self-esteem, social skills, decision-making autonomy, also known as "soft skills"*). Indeed, not only must one learn and acquire knowledge, but also learn to cooperate in order to work in a group according to the specific modalities of social networks: to show oneself on the network, to know how to manage one's visibility and to organize one's interactions and presence (Cardon 2008). Professionals are then enjoined to show flexibility and reactivity by mobilizing their social capital (Lazega 2006) via DESN functionalities that promote social and interactive (*interacting, posting a comment*) and cognitive (*problem solving*) dimensions in mediatized collaborative environments (Détienne *et al.* 2010). The objective is to develop innovation and social links, to improve production quality and to capitalize knowledge (Galinon-Mélénec 2010; Conaldi and Lomi 2013).

Finally, the DESN is more than just a technology. It becomes a collaborative partner in the service of value creation that stems from all the networking and co-configuration mechanisms in which professionals can participate. However, from the point of view of work activity, these mechanisms pose major challenges for professionals. We are witnessing a loss of the centrality of activities and a reconfiguration of the conditions surrounding the power to act.

2.3. Stakes for the development of activity: knowing how to give meaning to a poly-contextual and multi-mediated activity

Poly-contextuality is defined by a multiplication of socio-organizational (several types of organization within a company), professional (several professions) and instrumental (several means of interaction with DESNs and several mobilized ICTs) contexts. It is the main characteristic of contemporary activities carried out via these socially oriented technologies. In these variable, labile and flexible activities, the possibility of identifying "*what must be done, how and for whom*" participates in this sense of usefulness that is considered a priority in the construction of meaning in work. We can then wonder about the ability, of a professional immersed in these poly-contextual activities, to clearly identify the work objects that need to be carried out in order to orient the activity in these work environments and give meaning to their work (Sannino *et al.* 2009).

2.3.1. *Networking, the power to act and meaning at work*

2.3.1.1. *Activity, an organized and dynamic system*

In the field of work, the development of a sufficiently good activity is a health operator (Clot 2008). Human activity is characterized by the capacity of individuals and collectives to develop strategies with artifacts (to appropriate them) in order to be able to act on the environment around them. Work activity can therefore be considered as the result of a system of interactions produced between several human and non-human actors (artifacts, norms, rules, tools, technical objects) and which makes it possible to achieve objectives (see Figure 2.1).

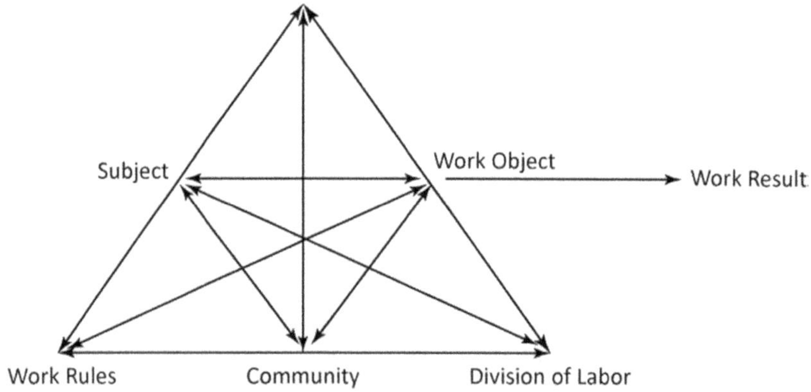

Figure 2.1. *The activity system (Engeström 1998)*

The subject will therefore configure and adapt their environment (material and technical) and its uses in new practices so that they correspond to their needs, intentions and motivations (Rabardel 1995). The activity must be considered as the result of the organization and permanent renegotiation that the actor produces, between these different extremes, to give meaning to their activity. According to this model, the individual can only carry out their action through systems of collective activity, where skills are developed, supported by practices validated by the group. We then speak of two mechanisms that are induced by mediatized activity: taking ownership of changes that come from outside (i.e. *learning*) (Rabardel 2005), and being able to act within the environment with the help of acquired knowledge (i.e. *developing skills*) (Bobillier Chaumon and Clot 2016). Thus, the power to act is defined as the effective range of actions that the individual or groups are capable of implementing, given a certain degree of re-creation of the environment (Bobillier Chaumon 2017). It also makes it possible to qualify the efficiency and dynamism of the activity.

2.3.1.2. *Network activity, the reign of negotiation in work*

Professionals using DESNs engage in a poly-contextual activity where different activity streams are managed in parallel, based on different work objects that may be juxtaposed or diametrically opposed (Engeström *et al.* 1995). The goals and means to carry out this activity are systematically different, involving collectives that often do not belong to the same profession, for purposes that may vary according to the people involved and the service objectives (Engeström 2008a).

The actor thus combines several activity systems that are constantly being impacted by the intentions and expectations of their colleagues, the effects of the division of labor and the tools used, as well as their own motivations and needs. The complexity is then to manage to articulate a professional practice that makes sense by "*knitting*" thier daily activity at the intersection of several activity systems and networks. This "*knitting*" makes it possible to formalize this extremely rapid, flexible and efficient mode of action ("*knots*"), which frees itself from the usual spaces and rules of operation, to focus on the moment and concrete needs. These "*knots*" correspond to the orchestration of a collective, distributed and partially improvised performance between actors or systems of activity that are otherwise weakly connected to each other (Engeström 2008a) (see Figure 2.2).

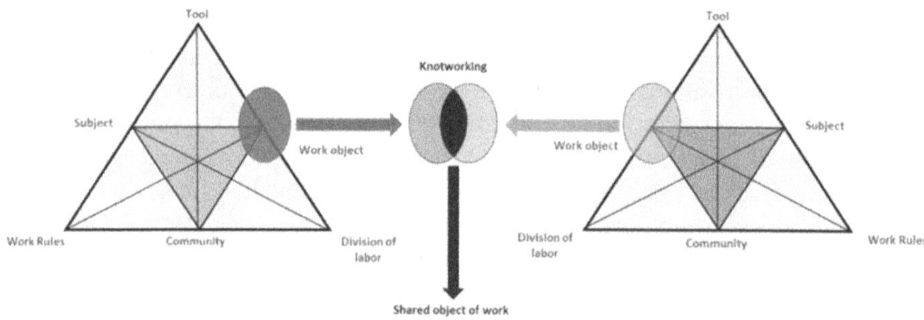

Figure 2.2. *Third generation of the activity system. For a color version of this figure, see www.iste.co.uk/bobillierchaumon/digital.zip*

2.3.2. *Sharing a degree of collaborative intentionality in a multi-mediated situation, a skill in its own right*

Shared collaborative intentionality helps us to design how professionals can engage in future collaboration, based on an assessment of how they anticipate, measure and evaluate commitment to investment (Engeström 2004; Edwards 2008). Derived from the criteria of activity theory, this concept is very interesting when defining asynchronous collaborations because it is based on an evaluation of the degree of shared collaborative intentions in a situation, that is, a sufficient degree of shared awareness of the actions to be carried out in a future collaboration (Engeström 2006).

The poly-contextual situations induced by the use of DESNs have a preponderant impact on collaborative dynamics (Engeström *et al.* 1995) by profoundly modifying and reconfiguring the way the activity is structured, which becomes more fluid (*regularly reconfigured*), partial (*may concern only certain aspects of the object of the activity*) and temporary (*may concern only a relatively short period of time*) (Engeström and Smith 2010). Sufficiently sharing collaborative intentionality with peers and via the system then becomes a means for the professional to adjust and quantify their commitment by weighing, through an analysis of the contingencies of the activity, the potential for collaboration expressed and the traces left on the system and the benefits and costs of their investment in collaboration (Harry *et al.* 2009; Edwards 2011). The professional, capable of integrating a sufficient understanding of the sharing of collaborative intentionality, is able to anticipate future activity (individual and collective) by identifying whether or not a work object is shared, evaluating the conditions for carrying out the joint actions to be implemented, defining the possible synchronization of actions via the system, anticipating constraints to overcome them and linking these future actions with the history of their activity (Bobillier Chaumon 2013). In contemporary working conditions, having a shared collaborative intentionality could become a precursor to engaging in the activity from a developmental perspective.

2.4. Conclusion

Collaborative platforms and DESNs are not, strictly speaking, technological innovations, since they aggregate technologies that have been in existence since the late 1990s. However, the combination of technologies associated with possible organizational uses profoundly changes the rules of interactions, social codes and collaboration dynamics, which are more fluid and less constant over time.

The principles lauded in the context of the implementation of networks have a major impact on the organization of work, its context and its intelligibility for individuals by multiplying the standards, rules, interlocutors, tools and purposes of work. In this way, work activity loses its centrality under the influence of the mechanics of social networks. At the same time, it intensifies and densifies in terms of flows, volumes and contexts. This induces constant changes in the structuring of the activity and in its organization over time (Engeström 2008a). The asynchronous collaborations resulting from the use of DESNs therefore imply identifying and understanding the expectations of colleagues, knowing how to make various points of view converge and to implement the methods for working collectively and being able to inscribe these actions, however fragmented they may be, in the overall history of individual and collective activity. Under these conditions, the power to act is possible because it reflects the ability of professionals to master this new collaborative environment with a certain degree of re-creation of the situation.

Giving meaning to work in these poly-contextual and multi-mediated situations therefore requires professionals to develop new skills for evaluating and anticipating future collaborative activity. In our opinion, it relies on the ability of professionals to identify a sufficient degree of shared collaborative intentionality.

2.5. References

Alter, N. (2009). *Donné et prendre. La coopération en entreprise*. Éditions La Découverte, Paris.

Barville-Deromas, N. (2014). Le réseau social numérique... Mais pour quoi faire ? Cas d'une non acceptation technologique. *Actes du 18ème Congrès International de Psychologie du Travail de Langue Française,* Florence, Italy.

Bobillier Chaumon, M.-E. (2013). Conditions d'usage et facteurs d'acceptation des technologies dans l'activité : questions et perspectives pour la psychologie du travail. Paper, Université Pierre-Mendès France, Grenoble [Online]. Available: http://recherche.univ-lyon2.fr/greps/IMG/pdf/HDR-Bobillier-Nov2013.pdf.

Bobillier Chaumon, M.-E. (2017). Du rôle des TIC dans la transformation digitale de l'activité et de la santé au travail. *La revue des conditions de travail*, 6, 16–25.

Bobillier Chaumon, M.-E. and Clot, Y. (2016). Clinique de l'usage : les artefacts technologiques comme développement de l'activité : synthèse introductive au dossier. *Activites*, 13(2).

Bourdieu, P. (1980). Le capital social. *Actes de la Recherche en Sciences Sociales*, 31(1), 2–3 [Online]. Available: http://www.persee.fr/doc/arss_0335-5322_1980_num_31_1_2069.

Cardon, D. (2008). Le design de la visibilité : un essai de cartographie du web 2.0. *Réseaux*, 152(6), 93.

Clot, Y. (2008). *Travail et pouvoir d'agir*. Presses Universitaires de France, Paris.

Cohendet, P., Créplet, F., and Dupouët, O. (2003). Innovation organisationnelle, communautés de pratique et communautés épistémiques : le cas de Linux. *Revue Française de Gestion*, 29(146), 99–121.

Conaldi, G. and Lomi, A. (2013). The dual network structure of organizational problem solving: A case study on open source software development. *Social Networks*, 35(2), 237–250.

Daniels, H., Edwards, A., Engeström, Y., Gallagher, T., and Ludvigsen, S.R. (2009). *Activity Theory in Practice. Promoting Learning Across Boundaries and Agencies*. Routledge, London.

Détienne, F., Barcellini, F., and Burkhardt, J.-M. (2010). La conception dans les communautés en ligne : questionnements thématiques et méthodologiques sur ces nouvelles pratiques. In *Ergonomie, conception de produits et services médiatisés*, Valléry, G., Le Port, M.-C., and Zouinar, M. (eds). Presses Universitaires de France, Paris.

Edwards, A. (2008). *Improving Inter-professional Collaborations: Learning to do Multi-agency Work*. Routledge, London [Online]. Available: http://search.ebscohost.com/login.aspx?direct=trueandscope=siteanddb=nlebkanddb=nlabkandAN=262359.

Edwards, A. (2011). Building common knowledge at the boundaries between professional practices: Relational agency and relational expertise in systems of distributed expertise. *International Journal of Educational Research*, 50(1), 33–39.

Edwards, A. (2012). The role of common knowledge in achieving collaboration across practices. *Learning, Culture and Social Interaction*, 1(1), 22–32.

Engelstätter, B. and Sarbu, M. (2013). Why adopt social enterprise software? Impacts and benefits. *Information Economics and Policy*, 25(3), 204–213.

Engeström, Y. (2004). Collaborative intentionality capital: Object-oriented interagency in multiorganizational fields. University of California, San Diego [Online]. Available: http://lchc.ucsd.edu/mca/Mail/xmcamail.2004_12.dir/att-0048/01-Collaborative_intentionality.pdf.

Engeström, Y. (2006). L'interagentivité orientée-objet : vers une compréhension de l'intentionnalité collective dans les activités distribuées. In *Sujet, activités, environnements : approches transverses*, Barbier, J.-M. and Durand, M. (eds). Presses Universitaires de France, Paris.

Engeström, Y. (2008a). *From Team to Knots*. Cambridge University Press, Cambridge.

Engeström, Y. (2008b). Quand le centre se dérobe : la notion de knotworking et ses promesses. *Sociologie du travail*, 50(3), 303–330.

Engeström, Y. and Smith, A.M. (2010). From teams to knots: Activity-theoretical studies of collaboration and learning at work. *European Management Journal*, 28(1), 80–82.

Engeström, Y., Engeström, R., and Kärkkäinnen, M. (1995). Polycontextuality and boundary crossing in expert cognition: Learning and problem solving in complex work activities. *Learning and Instruction*, 5, 319–336.

Galinon-Mélénec, B. (2010). Réseaux sociaux d'entreprise et DRH. Nouveaux outils pour de nouveaux enjeux ? *Communication & organisation*, 37, 13.

Lazega, E. (2006). Le capital social de l'organisation flexibilisée. *Revue française de gestion*, 32(163), 127–138.

Rabardel, P. (1995). Qu'est-ce qu'un instrument ? *Les dossiers de l'ingénierie éducative*, 19, 61–65 [Online]. Available: http://tecfalabs.unige.ch/mitic/system/files/rabardel_1995_quest-ce_quun_instrument.pdf.

Rabardel, P. (2005). Instrument, activité et développement du pouvoir d'agir. In *Entre connaissance et organisation : l'activité collective : l'entreprise face au défit de la connaissance*, Lorino, P. and Teulier, R. (eds). Éditions La Découverte, Paris.

Sannino, A., Daniels, H., and Gutierrez, K. (2009). *Learning and Expanding with Activity Theory*. Cambridge University Press, New York.

Teulier, R. and Lorino, P. (2005). *Entre connaissance et organisation : l'activité collective : l'entreprise face au défit de la connaissance*. Éditions La Découverte, Paris.

Zacklad, M. (2006). Documentarisation processes in documents for action (DofA): The status of annotations and associated cooperation technologies. *Computer Supported Cooperative Work*, 15(2–3), 205–228.

Zacklad, M. (2015). Genre de dispositifs de médiation numérique et régime de documentalité. *Les genres de documents dans les organisations, analyses theoriques et pratiques*, 145–183 [Online]. Available: https://www.academia.edu/30823408/Genre_de_dispositifs_de_médiation_numérique.

3

Virtual Reality: Definitions, Characteristics and Applications in the Workplace

In the context of the Factory of the Future, industries are implementing a number of technologies to improve their production, and to be more innovative and competitive. Virtual reality is one of these technologies. The objective of this chapter is to introduce this technology and explain how it can be used, both generally and in industries. We propose to use definitions of virtual reality that are commonly accepted by the scientific community. In this chapter, therefore, we define virtual reality from three angles: its purpose, its potential functionalities and its technical characteristics. We then present the various devices that constitute a virtual reality system: visual presentation devices, motion and position capture devices, proprioceptive and cutaneous feedback devices, and sound input and presentation devices. The final parts of the chapter are devoted to the applications of virtual reality. We begin with general applications (video games, art, tourism, education, vocational training, design, therapy, research). The last part of the chapter deals with applications of virtual reality in industry, giving examples of use for design and training.

3.1. Introduction

Industries are now entering a new industrial era, referred to as the Industry of the Future or the Factory of the Future. In this context, French companies are seeking, by increasing their level of dematerialization and implementing new technologies, to be more innovative, flexible and competitive. Virtual reality is one of these technologies. The Technologies Clés report (2016) by the *Direction Générale des Entreprises* identifies it as a technology of the future, alongside 46 other technologies, such as robotics, artificial intelligence or the Internet of Things.

Chapter written by Camille SAGNIER, Émilie LOUP-ESCANDE and Gérard VALLÉRY.

In view of its technical developments, virtual reality now seems easier to implement in companies, especially since head-mounted displays (HMD) are now accessible to the general public, their price has dropped sharply and many virtual reality applications are offered for sale.

The aim of this chapter is to better understand what virtual reality is, to present the type of devices it mobilizes, as well as to illustrate how it is used, particularly in professional contexts.

3.2. Some elements of definition

3.2.1. *The term "virtual reality"*

The first use of the term is attributed to Jaron Lanier, in 1989, who was then CEO of VLP Research Inc., a company providing computer equipment for virtual reality (gloves, glasses, etc.) (Steuer 1992). In English, the adjective "virtual" can be translated as "quasi, almost". The expression virtual reality therefore refers to an idea of quasi-reality.

The image given by the media tends to create a shortcut between the use of certain devices and virtual reality. Moreover, we imagine that virtual reality is limited to the use of a head-mounted display and a joystick. We will present definitions of virtual reality commonly accepted by the scientific community.

3.2.2. *The purpose of virtual reality*

The purpose of virtual reality may allow us to give a definition of it. For Fuchs *et al.* (2006), its purpose is "to enable one person (or several) to engage in sensory-motor and cognitive activity in an artificial, digitally created world that may be imaginary, symbolic or a simulation of some aspect of the real world" (p. 5).

User activity is described as sensory-motor and cognitive activity. We speak of sensory-motor activity because, on one hand, the user, in his/her interaction with virtual reality, perceives the virtual world and virtual entities through different senses (sight, hearing, proprioception, etc.) and, on the other hand, he/she acts *physically* within the world in which he/she is immersed. His/her activity is also cognitive, since he/she processes the information he/she perceives, memorizes it, makes decisions and undertakes actions. Note that there are perception–cognition feedback loops: a stimulus given by the computer part of virtual reality can lead to a motor action by the user, which in turn leads to new sensory stimuli that modify the user's intentions.

The second part of the definition mentions different types of virtual environments; they can be imaginary, symbolic or they can be simulations of certain aspects of the real world. Imaginary virtual environments are common in video games, where the aim is to immerse the player in a world different from their own. Other virtual environments represent symbolic worlds, in the sense that they materialize elements of the real world that are not always perceptible to humans. For example, they offer the possibility of manipulating DNA (Lamb *et al.* 2018). Finally, virtual reality is misrepresented as a pale copy of the real world, which it seeks to reproduce as closely as possible. However, the purpose of virtual reality, when it represents the physical world, is to exploit aspects of it for different purposes. For this reason, some virtual environments are considered to be "simulations of some aspects of the real world". For example, in a project review, a virtual environment can accurately represent the product you wish to evaluate, such as a machine. It is not necessarily interesting to represent the noises of the workshop, the routes of the parts supply trolleys or the neighboring stations in detail.

3.2.3. *A functional definition of virtual reality*

Another definition of virtual reality includes its potential functionalities. It allows us to: "extract oneself from physical reality to change virtually the time, place and/or type of interaction: interaction with an environment simulating reality or interaction with an imaginary or symbolic world" (Fuchs *et al.* 2006, p. 7).

Indeed, the use of virtual reality makes it possible to bypass the physical laws of our world. In virtual reality, the functioning of time is different; it is possible, for example, to stop time when you perform an action, or to go backwards in time. It also makes it possible to change place. While manipulating visualization and controlling devices in the physical world (imagine a user in an immersive room manipulating an arm with force feedback), he/she has the feeling of being in the virtual environment in which he/she is immersed (e.g. in an operating theater). Interaction with virtual reality is also different from interaction with the physical world, since it responds to the rules and workings of the application. For example, movements can be made by holding the button on a joystick rather than actually moving in the physical world.

3.2.4. *A technical definition of virtual reality*

We will end with a technical definition of virtual reality. It is: "a scientific and technical field exploiting computing (1) and behavioral interfaces (2) in order to simulate in a virtual world, the behavior of 3D entities, which interact in real time with each other and with one or more users, in pseudo-natural immersion (3) via sensor-motor channels" (Fuchs *et al.* 2006, p. 8).

Let us complete some aspects of this definition:

1) The term "computing" refers to all hardware and software parts of the system.

2) Behavioral interfaces are of three types:

– *sensory interfaces*, which provide information to the user about changes in the virtual environment. For example, a change in color may inform a user simulating an assembly operation, that the tool he/she is using has collided with another object;

– *motor interfaces*, which inform the system of the user's motor actions. For example, the system can exploit data on the user's position in an immersive room;

– *sensory-motor interfaces*, which provide information to both the computer system and the user. For example, a force feedback arm informs the system about the user's gesture, and at the same time forces the user to make efforts to simulate the resistance of the part being pierced.

3) Finally, we speak of "pseudo-natural" immersion, because the user does not act the same way in the virtual environment as he/she would *naturally* act in the physical world. There are sensor-motor biases in the interaction with virtual reality: for example, instead of walking from one point to another in the virtual environment, the user can teleport. Furthermore, the virtual environment does not necessarily provide all the sensory stimuli of the physical world.

It is also important to note that the definition of virtual reality is based on two conditions: the user must be immersed in a virtual environment, and he/she must also *interact* within this environment in real time. Therefore, 360° videos or 3D cinema are excluded from this definition, since the user is immersed in a world without really being able to interact with it.

3.3. The main interaction devices

Virtual reality systems are computer systems that include various peripherals. These are referred to as *devices* and can be classified in four categories (Burkhardt 2003):

– display devices;

– motion and position capture devices;

– proprioceptive and cutaneous feedback devices;

– sound input and presentation devices.

Each of these device categories will be discussed in the following sections.

3.3.1. *Display devices*

Visual presentation devices are the most common. It is rare to find systems that do not mobilize vision, although we could cite systems that were developed for visually impaired people that can incorporate spatialized sounds.

Virtual reality systems can be classified into three categories based on the degree of immersion of their visual presentation devices (Mujber *et al.* 2004). *Non-immersive* (or deskop-VR) systems are systems consisting of conventional computer monitors, which do not use specific hardware. *Semi-immersive systems* refer to widescreen displays, wall projection systems, interactive tables and head-mounted displays without stereoscopic vision. *Stereoscopic vision* is when the user has 3D vision of the virtual environment because a different image is displayed for each eye. Finally, *fully immersive systems* include head-mounted displays with stereoscopic vision and CAVE-type immersive rooms. In the latter, the virtual environment is viewed by users in cubic rooms, where three to six sides are screens.

In summary, non-immersive systems do not include specific technologies and do not offer stereoscopic vision, whereas semi-immersive and immersive systems use to specific technologies and can offer users stereoscopic vision.

3.3.2. *Motion and position capture devices*

Motion and position capture devices are also used. They provide the system with real-time information on the user's actions and position, as well as their evolution. Sensors can locate the user's entire body, a part of their body or an object held by the user (Fuchs and Mathieu 2011). Different types of sensors exist: mechanical sensors, electromagnetic sensors and optical sensors.

These sensors allow the virtual environment to react to the user's actions. For example, the system can detect the user's head movements and adapt the virtual environment accordingly, as if the user were actually in a place and turned their head to explore it visually.

3.3.3. *Proprioceptive and cutaneous feedback devices*

There are three types of proprioceptive and cutaneous feedback devices. The first are *haptic* (or *force feedback*) *devices*. They give information to the system about the movements made by the user, at the exact same time that they act on those movements. They are called *sensory-motor interfaces*. They allow the user to feel the stiffness, weight, inertia and friction of the objects they are manipulating (Gosselin and Andriot 2006). These devices include force-feedback arms, as well as wearable devices such as gloves and exoskeletons. For example, in industry training, these devices can transmit force to the user's arms to make the user feel as if he or she has to exert strong pressure on a workpiece in order to pierce it.

The second proprioceptive and cutaneous feedback devices are *tactile devices*. They provide sensory information to the user about contact with a virtual object. This information may concern its shape, roughness, texture or temperature (Benali-Koudja and Hafez 2006). These devices use matrix printer technology and Braille display systems for the blind. This type of device is still underdeveloped today.

The final proprioceptive and cutaneous feedback devices include *motion simulation devices*. Motion simulation devices place force on the user's body to change its orientation in space, or to cause it to undergo accelerations (Fuchs 2006). They include seats and other single-user devices and simulation booths. In the latter, the virtual environment display devices are included and they are usually multi-user. These devices can be used to simulate the movements experienced by passengers inside a car cabin or on a boat.

3.3.4. *Sound input and presentation devices*

Sound input devices include voice recognition systems. They allow the user to interact with the system through voice commands that are recognized and processed by the computer.

Sound presentation devices include software and hardware interfaces that enable spatialized sound (Tsingos and Warusfel 2006). They allow the simulation of spatial location cues. In other words, the intensity of a sound will depend on the distance between the source of the sound and the user. For example, music emanating from a radio set in a virtual environment will be perceived less and less strongly as the user moves away from it.

3.4. The main areas of application of virtual reality

Virtual reality can find different fields of application; whether in everyday life or at work.

3.4.1. *Applications in everyday life*

The use of virtual reality for leisure is certainly the most known by the general public. It is used in particular for video games. The arrival of Oculus Rift on the market in 2016, and the falling prices of this type of technology, have made it easier for players to acquire head-mounted displays. They have also prompted video game publishers to offer new products. The games on offer range from free games that come with the head-mounted displays, such as The Lab – which comes with HTC-branded head-mounted displays – to games that are already familiar to gamers on other platforms, such as Fall Out 4.

Virtual reality is also used in the field of art. Some artists paint in virtual reality. Various exhibitions of this type of work have been organized, for example, in the United Kingdom at the Royal Academy in London, or in France as part of the virtual reality exhibition, Laval Virtual. In addition, virtual environments have been developed to introduce the public to artists such as Rembrandt (*Meeting Rembrandt: Master of Reality*) or Van Gogh (*The Night Cafe: A VR Tribute to Vincent Van Gogh*).

Tourism offers users the opportunity to visit destinations in immersive virtual environments. For example, they can visit the Lake District National Park in the United Kingdom (tom Dieck *et al.* 2018) or the Fontanelle Cemetery in Naples (Marasco *et al.* 2018) in this way.

3.4.2. *Applications in various professional contexts*

Virtual reality is also used in professional contexts. In education, it is frequently used as a pedagogical tool, particularly for learning science. For example, it can illustrate the expression of proteins in mammals (Makransky *et al.* 2019) or the composition of molecules (Merchant *et al.* 2012).

Virtual reality is more widely used in vocational training contexts. There are a multitude of examples in this field. Surgery training is common (Fang *et al.* 2014; Khan *et al.* 2014), as is maintenance training (Gomes De Sá and Zachmann 1999; Ganier *et al.* 2013; Gavish *et al.* 2013). Applications have also been developed for assembly training, for example, for truck chassis (Ordaz *et al.* 2015) or car doors (Langley *et al.* 2016). There are also applications designed for risk prevention training, for example, in the construction sector (Sacks *et al.* 2013) or for home helpers (Polivka *et al.* 2019).

Virtual reality is also found in the field of design. In this context, it is used to replace or complement a physical prototype of a future product. Designers use virtual reality to inspect the product, to manipulate it, to evaluate it and to make design decisions. This applies to both vehicle design (see, for example, Berg and Vance 2017) and microwave oven fronts (Bruno and Muzzupappa 2010). It is also used in architecture (Paes *et al.* 2017), urban planning (Maffei *et al.* 2015) and landscaping (Guo and Yang 2013). The use of virtual reality to design manufacturing stations is common (Aromaa and Väänänen 2016; Bernard *et al.* 2019; Sagnier 2019).

Virtual reality is also becoming an increasingly therapeutic environment. It allows professionals to expose their patients to certain situations in a controlled and risk-free environment. In this context, it is integrated into the treatment of post-traumatic stress (Moraes *et al.* 2016) or delusions of persecution (Freeman *et al.* 2016). In addition, it can be used as a rehabilitation tool, for example, to treat balance disorders caused by certain pathologies (Hsu *et al.* 2017), or to offer upper limb rehabilitation after a cardiovascular accident (Shin *et al.* 2014).

Finally, virtual reality is proving to be a research tool. It enables the visualization and manipulation of different types of data, for example microscopic data (Theart *et al.* 2017) or atmospheric data (Helbig *et al.* 2014). It also allows psychological researchers to study user behavior. To illustrate this use of virtual reality, we can cite a study by Cœugnet *et al.* (2018) on the use of vibrotactile watches as a means of navigating a city, or a study by Morrongiello *et al.* (2019) on the behavior of children when crossing the road.

3.5. Applications of virtual reality in industry

In industry, virtual reality can be applied in two ways: at the design and training levels (see Moreau *et al.* 2004; Mujber *et al.* 2004; Hamid *et al.* 2014).

For design in industry, virtual reality is mainly used for virtual prototyping, project reviews and ergonomic studies of workstations. Virtual prototyping consists of replacing or supplementing physical prototypes of products or workstations with virtual modeling (Mujber *et al.* 2004). It is then possible for designers to conduct simulations, evaluate the product and make modifications at a lower cost. For example, Bordegoni and Ferrise (2013) illustrate how a virtual prototype can be used in product evaluation. The system they use is multi-sensory. It includes a haptic arm, a projection screen with stereoscopic vision and speakers. The authors present the use of this system to represent a washing machine. Participants in the study had to follow a scenario to interact with the virtual prototype and the physical prototype of the product.

Virtual prototypes can also be used for project reviews. In this context, different professions meet around a virtual environment and make design decisions on the designed product. For example, in the aeronautics industry, project reviews bring together project managers and the future fitters who will assemble the aerostructure, around virtual prototypes of manufacturing workstations (Sagnier *et al.* 2019). Finally, it is also possible to use the virtual prototype to carry out ergonomic studies of the workstations. For example, Bernard *et al.* (2019) simulated helicopter maintenance operations, in order to collect data on the operators' postures, the biomechanical constraints to which they were subjected, or their mental load.

Virtual reality also allows for virtual assemblies, that is, simulating the assembly of parts of an object (Seth *et al.* 2011). In this way, we can decide on the assembly sequence of the different parts, determine the physical and mechanical characteristics of the parts, choose the tools to be used and identify factors that may affect the quality of the assembly (Mujber *et al.* 2004). For example, Jayraman *et al.* (1997) developed VADE (Virtual Assembly Design Environment), an immersive virtual environment for virtual assembly. The virtual reality system used consists of a head-mounted display and a data glove to detect the user's hand movements. The system integrates data from CAD (Computer Aided Design) software on tolerances, part location, part orientation, etc. In the virtual environment, the user assembles or disassembles the different parts of an object, perceives the collisions of the parts and can record a trajectory, or reject it.

As for the use of virtual reality in industry training, this mainly concerns maintenance and assembly. For example, the study by Abate *et al.* (2009) presents a head-mounted display application with a force feedback exoskeleton for maintenance training in the aerospace industry. In this study, users were required to perform eight routine maintenance tasks. Similarly, Stork *et al.* (2012) propose a semi-immersive system with a wide screen and natural interactions (to grasp objects with the hands) for assembly training in the automotive industry.

Virtual reality is also used to train operators in risk prevention. These applications are developing particularly in the construction sector. For example, Sacks *et al.* (2013) compared the effectiveness of safety training given by (1) a classroom trainer with a slide show with (2) training involving an immersive virtual environment used alternately by the trainer and the learners. The virtual reality system included a projection screen that allowed stereoscopic vision. The virtual environment represented a construction site. Learners were asked to share their general knowledge of the construction industry, identify risks in the situations presented and explain the behaviors to be adopted. In the same vein, the study by Zhao and Lucas (2015) presented a non-immersive virtual environment to sensitize workers to electrical risks. It allowed users to detect risks while navigating on a construction site. The virtual environment provided information to users on risks and on the behaviors to adopt.

3.6. Conclusion

Virtual reality is one of the technologies that is implemented today and that will continue to be implemented in the coming years in professional contexts. In this chapter, we have enlightened the reader on what virtual reality is and on the devices that virtual reality systems mobilize. We have also summarized its main applications, in general and more specifically in industry.

Most of the studies on the use of this technology, including the majority of those mentioned above, adopt technocentric approaches (or a Technology Driver approach, i.e. developing products that then seek to gain acceptance; Davies and Buisine 2017). Studies that adopt this approach describe only the technical characteristics of a virtual reality system or virtual environment, as well as their main objectives (training operators to identify risks, conducting studies on the interaction of a customer with a product, etc.). However, perhaps because the implementation of this type of technology in professional contexts is not democratized, or because they are only exceptionally integrated into employees' activities, few studies have been carried out on their impact on work. However, their development raises questions about their suitability for users (usefulness, usability, acceptance), the changes they may bring about in the content of activities, as well as the health and well-being of employees. For these reasons, a user and activity-centered (anthropocentric) approach must be deployed to co-construct acceptable modalities for the deployment and adoption of these new immersive work environments.

3.7. References

Abate, A.F., Guida, M., Leoncini, P., Nappi, M., and Ricciardi, S. (2009). A haptic-based approach to virtual training for aerospace industry. *Journal of Visual Languages and Computing*, 20(5), 318–325.

Aromaa, S. and Väänänen, K. (2016). Suitability of virtual prototypes to support human factors/ergonomics evaluation during the design. *Applied Ergonomics*, 56, 11–18.

Benali-Koudja, M. and Hafez, M. (2006). Les interfaces tactiles. In *Le traité de la réalité virtuelle*, Fuchs, P. and Moreau, G. (eds). Les Presses de l'École des Mines de Paris, Paris.

Berg, L.P. and Vance, J.M. (2017). Industry use of virtual reality in product design and manufacturing: A survey. *Virtual Reality*, 21(1), 1–17.

Bernard, F., Zare, M., Sagot, J.-C., and Paquin, R. (2019). Using digital and physical simulation to focus on human factors and ergonomics in aviation maintainability. *Human Factors*, 62(1), 37–54.

Bordegoni, M. and Ferrise, F. (2013). Designing interaction with consumer products in a multisensory virtual reality environment. *Virtual and Physical Prototyping*, 8(1), 51–64.

Bruno, F. and Muzzupappa, M. (2010). Product interface design: A participatory approach based on virtual reality. *International Journal of Human-computer Studies*, 68(5), 254–269.

Burkhardt, J.-M. (2003). Réalité virtuelle et ergonomie : quelques apports réciproques. *Le travail humain*, 66(1), 65–91.

Cœugnet, S. (2018). Helping older pedestrians navigate unknown environments through vibrotactile guidance instructions. *Transportation Research Part F: Traffic Psychology and Behaviour*, 58, 816–830.

Davies, M. and Buisine, S. (2017). La culture d'innovation dans les organisations françaises. *Technologie et Innovation*, 2, 1–12.

tom Dieck, D., tom Dieck, C., Jung, T., Moorhouse, N. (2018). Tourists' virtual reality adoption: An exploratory study from Lake District National Park. *Leisure Studies*, 37(4), 371–383.

Direction Générale des Entreprises (2016). Technologies clés 2020 : préparer l'industrie du futur [Online]. Available: https://www.entreprises.gouv.fr/politique-et-enjeux/etude-technologies-cles-2020.

Fang, T.Y. (2014). Evaluation of a haptics-based virtual reality temporal bone simulator for anatomy and surgery training. *Computer Methods and Programs in Biomedicine*, 113(2), 674–681.

Freeman, D. (2016). Virtual reality in the treatment of persecutory delusions: Randomised controlled experimental study testing how to reduce delusional conviction. *The British Journal of Psychiatry*, 209(1), 62–67.

Fuchs, P. (2006). Les interfaces à simulation de mouvement et les interfaces à simulation de climat. In *Le traité de la réalité virtuelle*, Fuchs, P. and Moreau, G. (eds). Les Presses de l'École des Mines de Paris, Paris.

Fuchs, P. and Mathieu, H. (2011). Location sensors. In *Virtual Reality: Concepts and Technologies*, Fuchs, P., Moreau, G., and Guitton, P. (eds). CRC Press, Boca Raton, Florida.

Ganier, F., Hoareau, C., and Devillers, F. (2013). Evaluation des performances et de la charge de travail induits par l'apprentissage de procédures de maintenance en environnement virtuel. *Le travail humain*, 76(4), 335–363.

Gavish, N. (2013). Evaluating virtual reality and augmented reality training for industrial maintenance and assembly tasks. *Interactive Learning Environments*, 23(6), 1–21.

Gomes De Sá, A. and Zachmann, G. (1999). Virtual reality as a tool for verification of assembly and maintenance processes. *Computers and Graphics*, 23(3), 389–403.

Gosselin, F. and Andriot, C. (2006). Les dispositifs matériels des interfaces à retour d'effort. In *Le traité de la réalité virtuelle*, Fuchs, P. and Moreau, G. (eds). Les Presses de l'École des Mines de Paris, Paris.

Guo, X. and Yang, G. (2013). Animating prairies simulation with shell method in real-time. *Journal of Software*, 8(12), 3166–3172.

Hamid, N.S.S., Aziz, F.A., and Azizi, A. (2014). Virtual reality applications in manufacturing system. *Proceedings of 2014 Science and Information Conference*, 1034–1037.

Helbig, C. (2014). Concept and workflow for 3D visualization of atmospheric data in a virtual reality environment for analytical approaches. *Environmental Earth Sciences*, 72(10), 3767–3780.

Hsu, S.Y. (2017). Three-dimensional, virtual reality vestibular rehabilitation for chronic imbalance problem caused by Ménière's disease: A pilot study. *Disability and Rehabilitation*, 39(16), 1601–1606.

Jayaramn, S., Connacher, H., and Lyons, K.W. (1997). Virtual reality assembly using virtual reality techniques. *Computer-aided Design*, 29(8), 575–584.

Khan, M.W. (2014). Laparoscopic skills maintenance: A randomized trial of virtual reality and box trainer simulators. *Journal of Surgical Education*, 71(1), 79–84.

Lamb, R. (2018). Comparison of virtual reality and hands on activities in science education via functional near infrared spectroscopy. *Computers and Education*, 124, 14–26.

Langley, A. (2016). Establishing the usability of a virtual training system for assembly operations within the automotive industry. *Human Factors and Ergonomics in Manufacturing and Service Industries*, 26(6), 667–679.

Maffei, L. (2015). On the validity of immersive virtual reality as a tool for multisensory evaluation of urban spaces. *Energy Procedia*, 78, 471–476.

Makransky, G., Terkildsen, T.S., and Mayer, R.E. (2019). Adding immersive virtual reality to a science lab simulation causes more presence but less learning. *Learning and Instruction*, 60, 225–236.

Marasco, A. (2018). Exploring the role of next-generation virtual technologies in destination marketing. *Journal of Destination Marketing and Management*, 9, 138–148.

Merchant, Z. (2012). The learner characteristics, features of desktop 3D virtual reality environments, and college chemistry instruction: A structural equation modeling analysis. *Computers and Education*, 59(2), 551–568.

Moraes, T.M., de Andrade, A.F., and Paiva, W.S. (2016). Virtual reality for the treatment of posttraumatic disorders. *Neuropsychiatric Disease and Treatment*, 12, 785–786.

Moreau, G., Fuchs, P., and Stergiopoulos, P. (2004). Applications of virtual reality in the manufacturing industry: From design review to ergonomic studies. *Mécanique and Industries*, 5(2), 171–179.

Morrongiello, B.A. (2019). Effects of peers on child pedestrian behaviors in a virtual traffic context. *Journal of Pediatric Psychology*, 44(4), 509–516.

Mujber, T.S., Szecsi, T., and Hashmi, M.S.J. (2004). Virtual reality applications in manufacturing process simulation. *Journal of Materials Processing Technology*, 155(1), 1834–1838.

Ordaz, N. (2015). Serious games and virtual simulator for automotive manufacturing education and training. *Procedia Computer Science*, 75, 267–274.

Paes, D., Arantes, E., and Irizarry, J. (2017). Immersive environment for improving the understanding of architectural 3D models: Comparing user spatial perception between immersive and traditional virtual reality systems. *Automation in Construction*, 84, 292–303.

Polivka, B.J. (2019). Efficacy and usability of a virtual simulation training system for health and safety hazards encountered by healthcare workers. *Games for Health Journal*, 8(2), 121–128.

Sacks, R., Perlman, A., and Barak, R. (2013). Construction safety training using immersive virtual reality. *Construction Management and Economics*, 31(9), 1005–1017.

Sagnier, C. (2019). Etude de l'acceptabilité de la réalité virtuelle dans l'industrie aéronautique. PhD Thesis, Université de Picardie Jules Verne, Amiens, France.

Sagnier, C., Loup-Escande, É., and Valléry, G. (2019). Etude de l'acceptation de la réalité virtuelle : une complémentarité entre perceptions des utilisateurs et utilisation réelle de la technologie. Speech, *10th EPIQUE Colloquium*, Lyon, France.

Seth, A., Vance, J.M., and Oliver, J.H. (2011). Virtual reality for assembly methods prototyping: A review. *Virtual Reality*, 15(1), 5–20.

Shin, J.H., Ryu, H., and Jang, S.H. (2014). A task-specific interactive game-based virtual reality rehabilitation system for patients with stroke: A usability test and two clinical experiments. *Journal of NeuroEngineering and Rehabilitation*, 11(1), 1–10.

Stork, A. (2012). Enabling virtual assembly training in and beyond the automotive industry. *Proceedings of the 2012 18th International Conference on Virtual Systems and Multimedia, VSMM 2012: Virtual Systems in the Information Society*, 347–352.

Theart, R.P., Loos, B., and Niesler, T.R. (2017). Virtual reality assisted microscopy data visualization and colocalization analysis. *BMC Bioinformatics*, 18(64), 1–16.

Tsingos, N. and Warusfel, O. (2006). Dispositifs et interfaces de restitution sonore spatiale. In *Le traité de la réalité virtuelle*, Fuchs, P. and Moreau, G. (eds). Les Presses de l'École des Mines de Paris, Paris.

Zhao, D. and Lucas, J. (2015). Virtual reality simulation for construction safety promotion. *International Journal of Injury Control and Safety Promotion*, 22(1), 57–67.

4

Robotization in Industries: A Focus on SMEs

Based on a longitudinal case study conducted in a French industrial SME, this chapter aims to improve understanding of the challenges of robotization in small and medium enterprises. After the analysis of the implementation of a packaging robot in this SME, we will draw lessons from this technological adventure, which is still in a state of flux, around the need for pragmatic and rigorous support. In this respect, the Robot Start PME program (2013–2017) highlights the key factors of success of a technological change in the context of Industry 4.0, particularly in terms of methodological tools.

4.1. Introduction

According to Dudézert (2018), the notion of digital transformation has developed strongly in business discourse since 2013. Based on the argument of a quest for competitiveness, robotization, defined here as the implementation of robots to perform repetitive tasks in industry instead of humans, is being implemented at a fast rate around the world. In this chapter, "robots" refers to the industrial robots which are used to manufacture all types of products, although they are mainly present in the automotive industry. Welding, demolition, cleaning, packaging or monitoring robots are among the most common examples today. According to a June 2019 study by Oxford Economics, the number of robots used worldwide has tripled over the past two decades to reach a total of 2.25 million. Forecasts for 2030 predict around 20 million robots, including 14 million in China. According to the International Federation of Robotics (IFR), by the year 2018, 421,000 new robots had joined production lines, an

Chapter written by Sandrine BERGER-DOUCE.

increase of 10% compared to 2017. This trend is expected to continue with a forecast of, on average, +14% per year between 2019 and 2021. Estimated at a little more than 17 billion dollars, the industrial robot market remains largely driven by the automotive industry, whose major players are dragging behind them their subcontractors, who are mainly small and medium enterprises (SMEs). These small companies are often still perplexed by the stakes of robotization and, more generally, by the changes linked to the digital transition (IMT 2016). The question of the impacts on employment remains at the heart of business leaders' concerns (Conseil d'orientation pour l'emploi 2017). Globally, 20 million jobs, or 8.5% of the industrial workforce, could be replaced by robots by 2030 (Oxford Economics 2019). As a real competitiveness issue for organizations, digital transformation requires companies to adapt to the profound disruptive changes underway by using digital skills to create new business models, products and services. Intrinsically, this transition "leads to a real transformation of work organization: not so much because it allows all the organization's processes to be digitized, but above all because it requires each company to thoroughly rebuild a new balance between employee autonomy and control" (Dudézert 2018, p. 20).

The experiences of French SMEs with robotization still appear to be mixed. A study carried out in an SME specializing in packaging in 2017–2018, thanks to the financial support of the *Institut Mines-Télécom*, is symptomatic of a mixed technological adventure and invites us to rethink the support provided to these companies. Support does exist, however, but it remains ignored by the majority of SMEs. As an illustration, the Robot Start PME program, launched at the end of 2013 and supported by the French public authorities, aimed to guide 250 SMEs in their robotization projects. What lessons can be drawn from these insights, in terms of support for change induced by digital transformation in industrial SMEs? This is the objective of this chapter.

4.2. Focus on a robotization experiment in an industrial SME

4.2.1. *The case study in the service of exploratory research*

As this is an emerging topic, this study is based on an exploratory qualitative and inductive approach (Pratt 2009). A case study method was chosen to explore this questioning in greater depth (Eisenhardt 1989; Yin 2003). Since the case study did not require a representative sample of the population, we opted for a single case study of a longitudinal nature, from 2014 to 2018 (Yin 2003). The case study was constructed using a combination of primary and secondary data sources (Langley *et al.* 2013). Data collection was based on mixed methods, combining semi-directive interviews with management, department heads and workshop staff from April to November 2018, archival data, participation in an internal seminar with employee representatives (November 2018) and observations (site visits in 2014 and 2017) to examine the processes implemented during this period. The interviews and results were fully transcribed and analyzed, representing a corpus of more than 200 pages. We used a thematic analysis to process our data, which consisted of "systematically identifying, grouping and, subsidiarily, examining the themes covered in the corpus" (Paillé and Mucchielli 2012, p. 232).

4.2.2. *Presentation of the company*

Created in 1951, the SME under study has undergone several waves of consolidation with local companies, now belonging to a global group specializing in rigid and flexible packaging, owned by American investors since 2013. Spread over three industrial sites, the company had 320 employees in 2018, working in 3x8 shifts. Its industrial activity consists of the production of flexible packaging (extrusion technique) for mainly global customers (Unilever, Nestlé, etc.).

The digital transition in the SME was a strategic project supported by the company's management with the objective of "*taking care of all*" (Site Director). More concretely, the aim was (1) to reduce low value-added tasks, (2) to improve product quality, (3) to increase customer service and (4) to reduce psychosocial risks (PSR). Equipped with ERP (Enterprise Resource Planning) since 1998, the SME has embarked on a digital transition, more recently with the implementation of a production planning software in 2013 and, above all, the installation of a robotic arm, called Zoé, at the end of one of the plastic bag production lines in 2017, the explicit objective of which was to reduce the risks of occupational diseases while limiting tasks with very low added value. Several other projects have since been rolled out, in parallel, with heavy investments in the machine park in order to modernize it, most of the equipment being around 15 years old.

4.2.3. *A mixed technological adventure*

Following the installation of Zoé in a workshop in April 2017, then of quality control cameras in 2018, the motivation and recognition of employees was a subject of debate within the SME. Aware of these challenges, the Human Resources Department (HRD) created many initiatives in an attempt to anticipate the changes underway, particularly with regard to the employability of production staff. However, everything had to be invented within the company "by feel", to identify the professions (and therefore skills) of tomorrow. The HRD remained convinced that the *gestion prévisionnelle des emplois et des compétences* (GPEC, strategic planning of employment and skills) was the key tool to be mobilized in this context, drawing inspiration from the work of the 2017 *Conseil d'orientation pour l'emploi* (COE, French Employment Orientation Council).

Supporting employees in this digital shift is not obvious for the company, and initiatives are being tested along the way, with varying degrees of success. As an example, after Zoé's installation, in the spring of 2017, a short half-day training course was offered to a dozen production and maintenance operators who were then "released" to manage this new end of line assigned to the robotic arm. The result was a total fiasco, with repeated malfunctions, a very high level of non-quality, discouraging employees and a return on investment well below expectations (less than 50% after 12 months). This investment, which was supposed to reduce the arduousness of the tasks, to enable productivity gains (by the elimination of two temporary jobs) and to give a modern image of the SME, was finally the subject of mockery by many employees of the SME for several months: "The robot had bad press among the employees, it is a machine that takes the place of humans" (Methods and Maintenance Manager). In the end, three temporary workers were needed to control the work done by Zoé, which was the final straw! "In the space of 12 months, the robot had worked for a month…" (Point of contact technician). The management of the SME finally reacted by proposing, in July 2017, to appoint, as an internal point of contact, the technician who had been involved in the installation of Zoé, which he accepted, mainly out of personal curiosity. A few months later, he had surveillance cameras installed to detect the origin of malfunctions so that he could work closely with the robot's supplier to improve its operation. Despite this, the use of Zoé was totally suspended for several weeks at the end of 2017, due to unresolved technical problems.

4.2.4. *Lessons to be learned*

During a feedback meeting in July 2018, management acknowledged that the focus had been placed too strongly on the economic aspect of the robotization project (under pressure from American shareholders) to the detriment of the project's acceptability by operators and thoughtful integration upstream in the production line. As the point of contact technician explained, "Not all the guys on the shop floor knew […] we didn't know what the robot was going to do exactly […] it was installed, then production was asked to run it […]". From the managers' point of view, this experience was also a first for Zoé's supplier, by offering to work with flexible plastic. The learning curve for this actor, who had no experience in the field, was very steep. As for the workshop operators, they took a very critical position towards Zoé, even if they were, paradoxically, convinced that this was the future of their activity and were therefore waiting for other robots to be installed.

The SME analyzed appears to be typical of many traditional French manufacturing companies with "tunnel vision" management; a lack of resources, skills (especially digital skills), time and knowledge of support networks; a machine park that is partially very old and incompatible with connected digital technology. Nevertheless, the company was fully aware of the need to anticipate future changes despite the limited resources available. Faced with the aging demographic of the workforce, the major challenge was that of attracting and retaining talent, a recurring issue for industrial SMEs, further exacerbated in the specific context of the digital transition, due to the key skills to be gathered. Moreover, SMEs are highly dependent on their customers and suppliers (which are all very large companies), which unbalances the interplay of players in the value chain, as SMEs are strongly constrained by their external stakeholders, who are aware of their bargaining power.

Employee recognition is a sensitive subject from the point of view of both managers and employees. While everyone acknowledges the value of giving more autonomy to employees (especially in the workshops), opinions diverge on the incentives to be implemented: for example, technicians and operators with modest incomes are in favor of bonuses, while managers want to favor other non-financial levers. Beyond incentives, work organization needs to be reinvented in this context that is dominated by strategic paradoxes (Valenduc and Vendramin 2017).

In view of this experience, among the many avenues to be explored is that of accompanying the robotization project at a very early stage, to avoid being guided by technological euphoria. This supposes that we have identified existing initiatives on the market, which is not always obvious for industrial SMEs, which are, very often, still isolated.

4.3. Receiving support in order to better implement a robot: illustration by the Robot Start PME program

In 2013, the situation in France in terms of robotization was alarming: 33,000 robots installed, in comparison to 160,000 in Germany and 60,000 in Italy. Faced with this delay, which was considered detrimental to the competitiveness of French companies, public authorities decided to encourage primary robotization and to support 250 SMEs over five years. Thus was born the Robot Start PME program, which has since inspired the initiatives carried out in the context of the Industry of the Future in France.

4.3.1. *A program at the service of French industrial SMEs*

4.3.1.1. *Genesis, organization and results of the program*

Robot Start PME (www.robotstartpme.fr) is a national collective initiative financed to the tune of 33 million euros by the State, via BPI-France[1], as part of the future investment scheme, as well as by Symop (*Syndicat des machines et technologies de production*, French association for manufacturing technologies, a member of FIM, *Fédération des industries mécaniques*, Federation of mechanical engineering and metal working industries), Cetim (*Centre technique des industries mécaniques,* technical center for mechanical industry) and the *Institut CEA-List* (*Commissariat à l'énergie atomique et aux énergies alternatives,* alternative energies and atomic energy commission, Laboratory for Integration of Systems and Technology, one of three specialized technological research institutes of CEA Tech, the technological research arm of the CEA). This program offers, beyond financing 10% of robot investment, support from an expert chosen by Cetim. This support includes a one-day personalized diagnosis, a four-day feasibility and profitability study (functional analysis, skills engineering, Internet communication, economic model and validation of the specifications) and a five-day preliminary project, including choosing of the supplier (analysis of offers, meetings with potential integrators, formalization and follow-up of the financing file).

Of the 250 SMEs supported, 130 robots were finally installed at the end of the program in 2017. The beginnings of Robot Start PME were very encouraging, with 100 SMEs involved in the first year[2], of which more than 70% had less than 50 employees and 80% came from the manufacturing sector, then 170 at the end of the second year[3]. During the spring 2017 stage talks, observations were already contrasted between the enthusiasm of the official press releases of the consortium and the economic press. The program's stakeholders praised its ability to generate new dynamics that strengthened the strategic positioning of participating SMEs[4]. They based their comments on a survey conducted at the end of 2016 among the program's first 80 companies. The main results of this survey indicated a positive impact on economic development, an improvement in working conditions, the creation of new jobs and a strengthening of the strategic positioning of SMEs. In terms of economic development, productivity increased for 84% of the companies, and profitability for 74% of them, with an average growth of 15.8% in turnover and 27.5% in net profit in two years. In addition, 16% of the responding SMEs saw their international

1 BPI-France is a public investment bank, a French organization for company financing and development.

2 Press release from July 1, 2014.

3 Press release from March 24, 2015.

4 Press release from March 9, 2017.

competitiveness increase. The majority of the managers surveyed (89%) noted an improvement in their employees' working conditions, reflected in a reduction in physical hardship and an increase in skills through training, allowing them to move into supervisory or self-monitoring roles. Contrary to the fears that were still very present in companies, the arrival of the robot led to the creation of one to five jobs in 63% of SMEs, mainly in production and processes. In terms of strategic positioning, managers noted an impact on their business strategy and on innovation. As an example, brand image and trustworthy relationships with customers were frequently cited as benefits induced by the installation of the robot. As for the economic press, it seemed more cautious about the benefits of the Robot Start PME program. An article in the newspaper Les Echos[5], however, based on the press release from March 9, 2017, explained that this program "did not arouse the hoped-for enthusiasm among French SMEs" and that only 150 companies participated, out of the 250 initially targeted. The target of very small businesses did not seem to have been reached, the average size of participating companies that had integrated a robot in their workshops being 60 employees, and these SMEs were mostly located in three regions: Auvergne Rhône-Alpes, Pays de Loire and Brittany. The journalist then legitimately questioned the strength of adhesion to this program. The answer from the Symop General Delegate refers to a long-term need due to the complexity of the decision process related to the integration of a robot. It is often the entire organization of work in the SME that needs to be rethought, which implies supporting employees in new tasks and responsibilities. This is not an easy task in an aging industrial environment that is not attractive enough for the new generations.

The closing event of the program organized at the *Maison de la Mécanique* in Courbevoie on November 30, 2017 was the opportunity to unveil the results of a survey conducted among 106 beneficiary SMEs (29 of them using a questionnaire combined with interviews in the companies with the managers and employees) among the 130 that had installed their first robot thanks to the system. In terms of competitiveness, the increase in turnover was estimated at +18% in 2017 (compared to 2015) and that of the net profit at +55%. The arrival of the robot made it possible to increase the productivity of 86% of SMEs and profitability of 68%. As far as employment was concerned, 65% of companies had hired between one and five employees alongside robot integration. Overall, between 45 and 53 jobs were created in the program's SMEs and no jobs were lost. On the contrary, 84% of the companies surveyed planned to hire more new employees. The positive impacts on working conditions were notable for the vast majority of managers (90%). In concrete terms, the reduction in arduous working conditions (86% of managers), the increase in skills (71%) and, to a lesser extent, a new dynamic within teams (21%) were attracting attention. According to Aluminium Ferri's

5 Charbonnier, V. (2017). Les robots, une technologie complexe à intégrer pour les PME. *Les Echos*, April 5.

CEO[6], "this robot has created skills among many employees, especially for the operator who has invested in the project, which has gone from polisher to robot island pilot". Finally, the installation of the robot strengthened brand images (77% of respondents) and the relationships of trust with customers (61%).

4.3.1.2. *The example of* CFT Industrie

Among the SMEs benefiting from the program, *CFT Industrie* regularly reports its enthusiasm in the media and at professional events such as at the *Maison de la Mécanique* on November 30, 2017, on the occasion of the Robot Start PME review. Specialized in the bending and assembly of wires, tubes and sheets, *CFT Industrie* made the decision to invest in a welding robot. The installation of the robot, at the end of 2014, has literally transformed the organization of the company. Far from fears of job losses, the installation of the robot, costing 160,000 euros, allowed three new employees to be hired and, above all, a rise in team skills, even prompting the management to consider moving towards the liberated company… In a pragmatic way, the robot is synonymous with substantial productivity gains by dividing the welding time by three or four, depending on the products (a hospital trolley is now assembled in four minutes instead of the previous 15 minutes), while maintaining the same level of quality, without defects and with less scrap:

> This has enabled us to win new customers and diversify our business sectors. […] Our turnover has increased by 43% in four years. Today, it amounts to 2.5 million euros, including 150,000 euros in profits. *CFT Industrie* is now profitable[7].

For the operators, the robot has made it possible to move towards a reduction of musculoskeletal disorders (MSD), as welding operations done by hand can be particularly damaging. Working conditions have been significantly improved: a suction system reduces operators' exposure to welding gases and the ergonomics of the workstation have also been revised, thanks to the intervention of occupational health ergonomists. Moreover, the success of the project is due, in particular, to the upstream integration of operators:

> Two of the welders were appointed as project leaders. They went to meet the suppliers and worked on the specifications with Cetim. This involvement of the operators greatly facilitated the integration of the robot. Today, one of the two welders is in charge of its fine-tuning[8].

6 Garreau, M. (2017). Quatre PME témoignent des bénéfices obtenus après l'implantation d'un robot. *L'Usine Nouvelle*, December 4.

7 *L'Usine Nouvelle*, 3609, May 2, 2019, p. 60.

8 *L'Usine Nouvelle*, 3609, May 2, 2019, p. 60.

According to the supporting expert of *CFT Industrie*[9] (as well as from a dozen other participating SMEs), this SME's track record is exemplary in many ways. Having undergone a receivership in 2011, the company, which had 15 employees, has taken full advantage of the Robot Start PME program, which has enabled it to embark on a remarkable growth trajectory, by opening up to its ecosystem. For this expert, the critical success factors are the manager's personality, trust (in the expert, the project, the methodology and the integrator) and the shared vision of the robotization project. As explained by the company's management at the *Maison de la Mécanique* in late 2017:

> The investment goes far beyond the robot. There is everything we have worked on alongside it, in communication and management, in particular. [...] Today, each one of my employees has increased his or her skills [...] and the next step is the introduction of a cobot.

The SME now proudly displays its vision for 2025[10] in four areas: a 100% green company, a 100% competent and fulfilled company, a 100% connected and robotized company, and an international and locally successful company.

4.3.2. *An inspiring framework to support technological change*

The pragmatic methodology co-constructed within the framework of Robot Start PME is particular in that it is focused on the non-technological aspect of the industrial transformation, with only 20% a purely technological dimension. According to Pierre-Marie Gaillot (Head of the Robot Start PME project at Cetim)[11], the approach is resolutely of the "user-push" type, that is, inspired by the users themselves. As such, it adapts to all the technological elements identified by the *Alliance pour l'industrie du futur* (AIF) with which Cetim has been cooperating closely for years. Faced with the subject of robotization, which is often anxiety-provoking in companies, the decision to approach the subject from the angle of the acceptability of technological change and working conditions naturally led to links being established with specialists, in this case the ANACT (*Agence nationale pour l'amélioration des conditions de travail*, French national agency for the improvement of working conditions). This approach is in line with the current logic, which puts the emphasis on the human factor to succeed in the digital transformation in companies (Autissier *et al.* 2014; Agora Industrie 2018; Manuti and De Palma 2018).

For Robot Start PME, a network of 35 experts who were given the title of "*transformation architects*" was formed, in 2013–2014, to support SMEs in their

9 Interview from October 3, 2019, by telephone.

10 *Cetim Infos*, 245, March 2019, p. 17.

11 Interview from July 18, 2019, at Cetim.

digital transformation. Among them, 15 actively participated in the development of the methodology in a partnership approach. At the end of 2019, these experts constituted the core of a real community of practice, in the sense of Wenger (1998), available for other actions in the more general framework of the Industry of the Future. One of these experts explains that, since this first experience, they have been carrying out “360° diagnoses of the Industry of the Future” for Cetim. This group of actors, buoyed by the open innovation mindset, is an operational melting pot for the diffusion of a responsible and humanized approach to emerging technologies. Over the years, they have notably developed a common language which enables them to communicate more easily and quickly.

While these efforts to mobilize in favor of the digital shift are legitimately worth emphasizing, it is worth bearing in mind that the French public authorities have pledged to support 10,000 SMEs moving towards the industry of the future over four years. However, we must compare this ambitious figure of 10,000 with the 213,000 manufacturing companies nationally. This can explain the imperative need to further accelerate the movement to enable French SMEs to remain competitive on the international scene. For many of them, it is a question of survival!

> Between 2000 and 2015, the French manufacturing industry lost a quarter of its workforce, i.e. 820,000 jobs. Today, it accounts for 11.3% of market value added in France, compared with 13.3% in Spain, 15.8% in Italy and 22.6% in Germany[12].

4.4. Conclusion

One lesson learned from the Robot Start PME program is the need to take into account the long lead time in the robotization process in companies. Between the initial idea and the installation of the robot, five years have passed for the majority of the SMEs involved, including three years devoted exclusively to the maturation of the project. A study of the impact of the *Industrie du Futur* programs carried out by Cetim at the end of 2018 indicates that the deployment of tools and methods remains laborious, taking over two years in the majority of the supported companies. This shows, the importance of integrating the human factor in these reflections from the very beginning, to ensure that the induced growth will be guaranteed peacefully (Sousa and Rocha 2019). Several participating SMEs are now engaged in a development trajectory that is resolutely inscribed in the Industry of the Future. This is notably the case of ABCM[13], a family-owned SME with 75 employees, which today belongs to the *Vitrines de l'Industrie du Futur* and was awarded a prize at the

12 Cetim conference, June 13, 2019 – *Réussir la FrenchFab avec les PME : les propositions du Cetim pour accélérer et changer d'échelle*.

13 Telephone interview from October 3, 2019, with the expert supporting the SME.

Usine Extraordinaire organized at the Grand Palais in Paris at the end of 2018. As Pierre-Marie Gaillot explains:

> Each step calls for the next. The key to success is to set oneself in motion, to dare, to try [...] while having a clearly identified goal to give meaning to actions. The doorway chosen is less important than the act of taking a step.

A next step for SMEs is collaborative robotics. *Cobots* are there to work intelligently with humans on operations where they can interact safely, not to replace them.

The question of proximity to the actors seems crucial in guaranteeing the success of the support to achieve digital transformation. This shows us the importance of continuing to network the territory to increase this proximity that is essential to the dissemination of support offers to isolated SMEs (those outside the network, for example). This is because in the plans, in the end, we always find the same SMEs.

According to the French Ministry of the Economy, approximately 185,000 companies are likely to be sold over the next five years in France each year. The buyers of these companies, who are more sensitive to human issues and their employees' search for meaning in their work, could play the role of translators, concerning digital transformation. In other words, research on business takeovers as a catalyst for digitalization would undoubtedly deserve the full attention of our academic communities. However, the road to robotization still seems long for French companies, if we consider the density of 132 robots per 10,000 employees in 2015, compared to 309 in Germany, the most automated country in Europe, and 185 in Italy, according to statistics from the International Federation of Robotics. The patchwork of the financial aid to which French SMEs are entitled is not really designed in a way to make their job easier, as the number of different schemes has mushroomed in recent years. Finally, assessments of programs dedicated to the digital transformation of SMEs report that investments are largely focused on production equipment to the detriment of investments directly related to digital technology. It's hard to know where to start…

4.5. References

Agora Industrie (2018). Révolution humaine ? Un nouveau rôle pour les hommes et les femmes de l'industrie du futur. White paper [Online]. Available at: http://www.industrie-dufutur.org/Documents%20%C3%A0%20t%C3%A9l%C3%A9charger/agora-industrie-livre-blanc-2018-revolution-humaine-nouveau-role-hommes-femmes-de-lindustrie-futur/.

Autissier, D.J., Johnson, K., and Moutot, J. (2014). La conduite du changement pour et avec les technologies digitales. *Question(s) de management*, 7(3), 79–89.

Conseil d'orientation pour l'emploi (2017). Automatisation, numérisation et emploi. Report, Conseil d'orientation pour l'emploi, France.

Dudézert, A. (2018). *La transformation digitale des entreprises*. La Découverte, Paris.

Eisenhardt, K.M. (1989). Building theories from case study research. *Academy of Management Review*, 14(4), 532–550.

IMT (2016). Entreprise du futur : les enjeux de la transformation numérique. White paper [Online]. Available: https://www.imt.fr/wp-content/uploads/2017/07/imtlivreblancvfweb-161209104735.pdf.

Langley, A., Smallman, C., Tsoukas, H., and Van de Ven, A.H. (2013). Process studies of change in organization and management: Unveiling temporality, activity, and flow. *Academy of Management Journal*, 56(1), 1–13.

Manuti, A. and De Palma, P.D. (2018). *Digital HR: A Critical Management Approach to the Digitalization of Organizations*. Palgrave Macmillan, London.

Oxford Economics (2019). How robots change the world – What automation really means for jobs and productivity. Report, Oxford Economics, Oxford.

Paillé, P. and Mucchielli, A. (2012). *L'analyse qualitative en sciences humaines et sociales*. Armand Colin, Paris.

Pratt, M.G. (2009). For the lack of a boilerplate: Tips on writing up (and reviewing) qualitative research. *Academy of Management Journal*, 52(5), 856–862.

Sousa, M.J. and Rocha, A. (2019). Digital learning: Developing skills for digital transformation of organizations. *Future Generation Computer Systems*, 91, 327–334.

Valenduc, G. and Vendramin, P. (2017). Digitalisation, between disruption and evolution. *Transfer: European Review of Labour and Research*, 23(2), 121–134.

Wenger, E. (1998). *Communities of Practice – Learning, Meaning and Identity*. Cambridge University Press, Cambridge.

5

Serious Games for Vocational Training: From Emotional Labor to Knowledge Transfer

Based on research on a serious game used for training purposes, this chapter aims to analyze knowledge transfer related to emotional labor between gaming and work situations. After defining emotional labor, we will discuss issues related to the use of a serious game as mediation; whether or not it is accepted; and whether it contributes to the possibility of better regulating emotions. The results suggest that knowledge transfer, in and around work situations, is related more to the acceptance of mediation, including its debriefing, than to the playful attitude itself.

5.1. Introduction

Among the many definitions of serious games, we refer to the definition from video game designers Chen and Michael (2005): "A serious game is a game that is designed for purposes other than pure entertainment." According to this definition, a serious game can apply to a large number of sectors: education, health, advertising, communication, politics, humanitarian, defense, religion, art, etc. – serious games can be digital or non-digital (Abt 1987). To specify the digital form, the term "serious video game" may be used.

Alvarez *et al.* (2016) define the serious game as a: device, digital or non-digital, whose initial intention is to combine, with consistency, both utilitarian ("serious")

Chapter written by Lydia MARTIN, Julian ALVAREZ and Antoine TALY.

aspects such as, in a non-exhaustive and non-exclusive way, teaching, learning, communication, or information, with playful motives (competition/challenge, rules of control, closure and procedure, scoring, the artificial character) stemming from the (video) game. Such an association is aimed at an activity or a market that goes beyond mere entertainment (p. 17).

The "utilitarian scenario" refers to the work of researcher Étienne Armand Amato, who in 2007 proposed the term "utilitarian game" to designate serious games. For Amato, "utilitarian" is understood as aiming to: [...] bring about a transformation in their addressees, in the sense of an improvement in skills (training), adaptation to an environment (treatment of phobias), understanding of a phenomenon (education) or greater adherence to the message conveyed (promotion, advertising, ideological video games, also known as political games) (Amato 2007). Three main families of utilitarian functions, to be associated with the game, can be referenced: to broadcast messages, to train and to exchange data (Alvarez *et al.* 2016).

Our research focused on a serious escape game (SEG). The principle of escape games (EG) is simple: players are in a room and have to get out in a limited amount of time (usually an hour). Whether the room is real, in a live action escape game, or virtual, in board games or video EG, does not change this principle. Used by companies as entertainment in team-building seminars, EG has become a new type of serious escape game (SEG), and is starting to be used in training and recruitment schemes as well.

A French company has created a SEG dedicated to apprenticeships. The game is in a case divided into three boxes, each closed with a padlock and all the clues are recorded on a tablet. A company in the insurance industry wanted to adapt this SEG so that their employees could discuss the emotional charge of customer relations, which is a key element of their activity, in a playful way. A major problem in customer relations is the emotional response to the various situations encountered at work. Understanding and regulating emotions can be a major asset in customer relations to manage incivilities and difficult situations, and make appropriate and professional decisions. The training aims to help professionals take a step back during difficult exchanges and to help them regulate their emotions. This implies that professionals need to know and master strategies to regulate their emotions. This is called "emotional labor".

There is tension between the daily life of customer advisers in the field and the "frivolity" of gaming. Therefore, it remains to be determined whether games, and in particular the proposed SEG, are an appropriate mediation to address such issues with often overwhelmed professionals. Is the artifact used considered and experienced as a game? Does the training session produce a change in the behavior of the professionals at work? We proposed to address these questions through action research, within the framework of the psychosociology of work (Lhuilier 2013). The group interviews were analyzed using two methods:

– *in situ* observations made during gaming activities;

– questionnaires submitted one month later.

The data studied analyzed issues related to: whether or not an SEG is used as a mediator; whether or not such mediation is accepted; and whether or not it contributes to the possibility of a better regulation of emotions. Through this process, it is possible to analyze learning transfer (Frenay and Bedard 2011) in work situations.

5.2. Emotions, debriefing and learning

5.2.1. *Emotional labor*

To define emotional labor, we must distinguish between affects, emotions and feelings (Martin and Alvarez 2018). Shouse (2005) explains the difference between "feelings, emotions and affects": affects are abstract and unconscious, feelings are personal and biographical, and emotions are social. An affect is an intense, unconscious experience. It is highly subjective, as affects cannot be fully expressed through language, they are unconscious (Massumi 1987; Shouse 2005). A feeling is a personal sensation related to prior experiences and life history, as each person has a distinct set of prior sensations which they draw from, in order to interpret and categorize their feelings. An emotion would be the "showcase" of a feeling. An emotion can be an expression of our internal state or can be adapted to meet social expectations.

Social rules impose the need to adapt our emotions, what the American sociologist Hochschild (2003, 2017) calls "emotional labor". This work allows professionals to adapt and regulate their emotions in order to make them "appropriate" to the situation and context. In some professions, individuals need to adapt their emotions in order to adapt them to society's expectations. Emotional labor is the expression of emotions in the service of work. For example, a flight attendant must be smiling and welcoming and repress anger, indifference or worry in order to achieve "well-being" and recognition from passengers. Emotional labor involves changing the emotional state of employees through a variety of techniques and strategies. As a result, there may be a dissonance between what the person is feeling (affect/feeling) and the emotion they need to show, and this emotional labor can have a long-term cost on health. According to Hochschild, emotional labor is characterized by "transmutation", that is, a transfer of emotions from the private to the public. Emotions, which are normally in the private domain, are managed and marketed by the company in the public space. Lhuilier (2006) studied this evolution within organizations, that ranged from the proscription to the prescription of emotions at work. Emotional labor is the most damaging to health when the emotions prescribed at work do not accurately represent the experience and feelings experienced by the individual. Emotional dissonance among employees is strong when a customer is perceived as aggressive and harassing. In this case, it is more difficult to express positive emotions with an unpleasant and aggressive client (Ashforth and Tomiuk 2003). Zapf (2002) demonstrated the importance of social support, including colleagues, in preventing dissonance.

Customer advisors are faced with emotional labor that can be difficult without having the theoretical background to understand it. The objective of the training is therefore to transfer this knowledge and to enable professionals to use it in their daily work. The learning is discussed during the debriefing to encourage knowledge transfer to work situations.

5.2.2. *Learning in a game*

With respect to learning through games, studies conducted by Sherry Turkle in 1986, showed the acquisition of skills through Pac-Man (Turkle 1986). More recently, Rebetez and Bétrancourt (2007), albeit cautiously, listed a series of studies devoted to the psychological and cognitive contributions of video games: for example, improving performance related to mental processes by playing Tetris created by Pajitnov in 1984, or improving performance related to dynamic spatial reasoning by playing Marble Madness (Rebetez and Bétrancourt 2007). Yann Leroux, for his part, identified a series of skills that video games could develop in terms of clinical psychology: controlling anxiety, establishing social contacts, integrating personality, etc. (Leroux 2012, pp. 74–78).

This non-exhaustive list provides examples of how games could support different types of learning. In the workplace, it can also make sense to use games to provide training and for companies to invest in training to encourage the learning of expected "behavioral skills". In particular, games can facilitate cooperation within teams (Schmidt 1990; Taurisson and Chunikin 2005), developing the ability to negotiate, discuss, collaborate and share emotions and ideas with others (Sauvé *et al.* 2007). Since SEGs are team games, it is tempting to use them to promote or evaluate teamwork. Two studies suggest that it is possible to develop group dynamics in the context of an EG (Pan 2017; Warmelink 2017).

In our research, the question was to analyze whether games, debriefing and associated work groups promoted the transfer of learning in work situations.

5.3. The context and framework of the Serious Escape Game (SEG)

5.3.1. *Difficulty in gamification*

Customer relations imply performance, customer satisfaction and objectives responding to requirements in a complex and difficult relational environment. The training in our study aimed to:

– make teams aware of the roles of emotions and their individual and collective regulation in customer relations;

– provide operational means to develop the emotional competence of teams and manage work situations differently;

– support managers in the implementation of the proposed system in terms of customer relationship issues.

The digital SEG thus made it possible to combine the theoretical concepts illustrated, with practical cases to facilitate their appropriation, strong moments of collective reflection and exchange of experiences for:

– sharing the difficulties encountered on the theme and identifying appropriate actions;

– enriching the training with the experience and knowledge of the teams on the theme;

– allowing each participant to discover not only external resources (within the group), but also internal resources (individual potential).

These resources were explored during the various challenges offered in the game.

The challenges embedded in the SEG were illustrated with concrete examples from authentic work situations. For each of the learning objectives (cognitive biases, incivilities and people in difficult situations), the work situations encountered by the teams were detailed. When designing the game, each theoretical part was associated with practical cases from these work situations.

5.3.2. *The training system*

The training program involved 50 customer advisers, but only 45 were able to attend. The program consisted of a full day with the digital SEG and complementary half-days for 10/12 professionals to work and take the work situations modeled in the game further, or any situation involving the adviser.

The day started with an introduction by the team leader, explaining the game choice and introducing the theme of emotions. The teams were randomly composed of three or four people, with the exception of the team of the three managers.

Once the game was introduced, each team could start playing. Only three participants had already played live action escape games. The game lasted from 1h15 to 1h30. Feedback from the game was recorded through the team's creation of drawings. Each participant was given a large sheet of A4 paper, and then teams of three to four people had to make a collective illustration of what happened during the SEG and indicate what they retained from the acquired knowledge to transfer to work situations.

In the afternoon, the videos made by the participants during the game were viewed. Another group, who did not make the video, reacted on the emotions the adviser could feel and how he/she could regulate them. Then, the day ended with a quiz on the day's learning. A memento containing the main knowledge from the training was distributed to each participant.

5.3.3. *A research-action approach*

We carried out action research that aimed for global and transformative change through the simultaneous process of action and research, the two being linked by critical reflection.

In this action research, we used a qualitative part, with observations during and after the game, as well as during debriefing, and working groups, a process quite similar to the analysis of practices. We supplemented our observations with a questionnaire. The objectives of the questionnaire were both to collect data on the experience of the game, to analyze the involvement of professionals in the game itself and in the role of debriefing and working groups learning.

5.4. Results

5.4.1. *Observations during gaming*

The 45 professionals were divided into 10 teams of four to five people during the SEG sessions. We could observe the surprise when they understood that "*it was time to play*". Some of them had played an EG a short time before and said they were very excited to play again. Most of the teams played in a good atmosphere and cooperated with each other. The competition between the teams came about relatively quickly. From our experience, the game seems to summon competition quickly. The teams evaluated each other by observing where the others were at in the game. The music from some of the puzzles was a very good indicator and we heard: "*What is that? We're not there yet*", or when their music rang out, they said, "*We're only there, they're ahead of us*". This performance evaluation mainly concerned teams that were close to each other. The 10 teams and the layout of the room did not allow them to progress in the game and follow the progress of the other teams at the same time. Two teams caught our attention.

The first team were not having fun; two of the three members spent a lot of time thinking about and discussing all the puzzles, and it was very difficult for them to get involved in the game. The third member was bored and fed up because he couldn't have fun. We discussed the reason for this difficulty in getting into the game with them and both members expressed their disagreement with playing.

The second team played and collaborated throughout the game. Towards the end, people started to get nervous. They couldn't find the material to answer the riddles. At first, we thought it was a logistical mistake but we realized that they had not opened the last box where all the puzzles were. When we talked to them, they explained that they were falling behind and that they wanted to go faster by just using the tablet. The fact was that this was not possible, the puzzles and the tablet worked together. The team's defense strategy was to say that the boxes had been badly prepared and the material they needed was missing, which stopped them from continuing, so they lost time. This behavior was similar to the behavior of the bad player who "destroyed" the game: they said "it is not a good game" when the results did not meet the team's expectations (Mehl 1981). In this example, the team tried to cheat, that is, they tried to go faster using only the tablet, although this was not possible.

These two examples are isolated, most of the teams played and expressed pleasure, in particular by making two videos to illustrate work situations related to the game. Playful behavior was analyzed when using the questionnaire sent to them after the game, the debriefing and the working group.

5.4.2. *Post-game discussions*

5.4.2.1. *The debriefing session*

After the "hot reactions" within each team, the second phase of the debriefing was carried out by creating a collective illustration within each team of what happened during the SEG and how people thought they would transfer the new knowledge to work situations.

The collective illustration was the result of the collective discussion of three questions:

1) How did you find this experience?

2) What do you think of the way the team worked?

3) How can you learn from this experience and transfer it to your professional activity?

The idea of a collective creation was to get the teams to discuss what happened during the SEG and what they understood about emotional labor, and make it visible. The drawings were in light colors and showed the fun and enjoyment of the teams during the SEG. The experience was also depicted with brains, hearts and people talking together. The transition to work situations was illustrated with arrows from learning to work situations. Only one team retrieved their drawing to share in their workplace.

5.4.2.2. *Working groups*

Two months after the SEG and its debriefing, working groups were organized to deepen emotions at work. We had four groups of 10 employees and one group of managers.

We started the session by asking three questions:

1) What do you remember about the training?

2) Do you have a different approach to situations since the training?

3) What are your expectations today?

For the first question, all players appreciated the discussions and interactions with colleagues in the game: "several discussions on work situations"; "good collaboration with colleagues"; "importance of each colleague's skills". Most of them remembered the main topic of the EG and some of the teachings on emotion regulation: "importance of managing our emotions"; "emotion is a very personal feeling and above all very different from one individual to another"; "don't confuse your emotions with the situation"; "maintain distance from the situation".

The EG was a good memory for all except one participant, a member of the team who was reluctant to play. Two participants were frustrated because they would have liked to go further in work situations and because of the competition. Going further was everyone's expectation for the work groups.

The question about the transfer of learning was very interesting. For some (one-third), the transfer of learning occurred in everyday life: "more relaxed in person or on the phone, more fluid conversations"; "since the escape game, I try to pause before any reaction"; "I am more aware of my emotions, my reactions and what I can feel"; "I take a step back from other people's emotions, I let myself be less invaded by others' emotions"; "I know better how to react to a person in a totally stressful situation, distress". I have learned that the person is unable to receive information until the brain is calm and soothed. Only then will we be able to guide and help him/her. For others (two-thirds), the transfer did not take place: "the learning was blurred"; "too far to remember the content"; "I didn't really change"; "the game took over the content".

The expectations for the working groups were: "to manage my feelings by maintaining distance, but not too much"; "to gain a better knowledge of my colleagues' feelings"; "to manage my emotions according to the situations"; "to understand how to better manage customers who irritate me"; "to be more detached from my daily work"; "to better manage what irritates me"; "to take a step back"; "to learn things, to put words to my feelings"; "to manage conflict situations". A few were looking for tools: "to find tools to manage conflicts or difficult situations on the phone". The difficulty was trying to make them understand that it is more important to understand an emotion than to banish it.

The second part of the working groups was devoted to explaining, for each of them, a work situation and the associated emotion, in order to analyze it together. The ANACT model for analyzing a problematic situation was used (Sanglerat *et al.* 2014). This model promotes the interaction between participants in the identification of psychosocial risks and solutions. This involves: describing problematic situations with the associated emotion, consequences, causes, risks and actions. Each person explained a work situation and the others asked questions.

After this analysis of the work situation, the professionals envisaged some changes: "maintaining discussions on work situations with colleagues"; "asking the question of causes: what makes the other person act/interact equally"; "analyzing emotions and taking a step back"; "choosing the right time to contact a client";

"asking a question at the end of the e-mail or conversation to encourage the answer"; "using key words: I am surprised … I need …".

From our observations, the majority of people were curious, while maintaining a professional attitude, despite the playful context of the training. A minority of 4/5 people, initially more reticent, resistant, assertive and very spontaneous, got involved and participated positively. Only one person remained on the sidelines and did not wish to participate without giving a reason. All the professionals felt the need to speak, and each had a personal anecdote to share. Some of them asked to continue the working groups.

5.4.3. *Questionnaires*

The questionnaires were completed by 25 of the 45 participants. The answers to the questionnaires were analyzed with the aim of examining: (1) the perception of the training session as an authentic game; (2) the transfer of learning, in particular through the debriefing session and the post-game working groups; and (3) the potential impact of the training session on professional practices.

Professionals were asked about their perception of training through a series of questions. The questionnaire included closed-ended questions about the gaming experience in general and this experience in particular. Most of the questions were developed based on the results of initial research (Martin 2018). The flow scale was used to subjectively assess the learners' enjoyment and motivation in gaming (Heutte *et al.* 2014). They were asked directly if they felt that SEG was a game. Almost all (24/25) responded that they thought it was a game.

The other questions were indirectly related to their perception of "the session as a game: (1) I felt that I had the ability to deal with the situation; (2) I did not see the time go by; (3) I felt constrained by the rules of the game, etc." These indicators were used to analyze whether each individual behaved as they would in a classic game situation (without the injunction to play): that is, a person who did not see time passing, who did not feel constrained and who felt in control of the situation was considered to be playing, whereas a person who felt constrained and did not lose the notion of time was considered to not be playing. With this approach, two people were considered as not playing and four were considered neutral.

Another set of questions analyzed their perception of training. One of the questions asked whether, in their opinion, SEG was appropriate to address the issue of emotional workload in their work. Another questioned the debriefing to allow for the transfer of learning into work situations. Finally, a few questions examined their behavioral change. One question asked if they used the memento handed out during the training. Another question asked whether they had succeeded in transferring the knowledge acquired during the training into their work. Although the sample size was limited, it appears that the three observations (gaming, transfer and effect) were not independent (Friedman's test, sum of the ranks, $p < 0.02$). Looking at the paired associations, we found no significant association between play and effect (Kruskal–Wallis test by ranks, $p>0.15$). This is consistent with the idea that play behavior does not appear to be crucial in maintaining the effect of play itself. On the other hand, we observed a significant association between transfer and effect (Kruskal–Wallis test by ranks, $p < 0.02$). This confirms the importance of debriefing, as suggested by others (Garris 2002). These results also demonstrate that the questionnaire is capable of capturing this association.

5.5. Discussion and conclusion

This study shows that the effect of training on the participant's work situations was significant, although we limit ourselves to those for which there was no ambiguity (11/25), although this should probably be pursued with larger cohorts.

One of our questions was whether it is possible to use a game for an activity that has a very strong emotional imprint, as with emotional labor. An interesting element is that although almost all of the participants defined the activity as a game, a significant fraction of them did not demonstrate play behavior. This was not surprising, given that the same game can have different meanings for different participants (Juul 2010). Indeed, a study of adult education using a serious game revealed the significant effect of participants' previous experiences as players; this preliminary knowledge altered the playful experience and perception of the serious game as an evaluation (Martin 2017, 2018). In addition, a small study of an SEG found a correlation between the gender of participants and the effect on their experience (the "flow") (Hou 2012).

The questionnaires were designed to capture information that is usually collected through observations or interviews. This should make it possible to collect data on larger cohorts in the future. However, the questionnaires showed less consistent responses, with related responses suggesting that some questions could be improved. It may also be that some indicators are more appropriate than others. Another limitation of this study is related to the fact that the questionnaires were anonymous, which prevents direct comparison with the results of observations and interviews.

Emotions are the visible part of an affect. Emotional labor can consist of regulating this visible part, but what about the affect? If the affect is the unconscious part of the emotion, the part we do not control, how can we regulate it without analyzing and understanding it? Individuals are asked to regulate the visible part, the emotion, from a deeper part, the affect. In the distinction between feeling, emotion and affect, we can assume that debriefing allows us to discuss the feeling and to think about how to regulate it according to the prescribed emotions, which is emotional labor.

It is an illusion to believe that playing is enough to promote learning. Debriefing is essential to establish links with work situations and thus promote knowledge transfer. The post-game working groups on the emotional load at work made it possible to deepen this link initiated during the debriefing. Our results are consistent with the notion that a game is a mediator and not the final solution, which made it possible to collectively discuss work situations during the debriefing.

In this example, the modification of feelings could be the "actual work" and the modification of emotions could be the "prescribed work". The effect will remain hidden or ignored. A tension may arise between the feeling, affect and emotion that can be harmful to health (Lhuilier 2006). Interestingly, most participants told us that the support of managers and co-workers was essential to discussing these daily tensions and to express sadness, anger, tears and all the emotions prohibited at work, as well as in relationships with customers. This is what participants said the post-game working groups fostered. Since social support prevents dissonance (Zapf 2002), it can reduce tension, because emotions are prescribed in front of consumers, not colleagues. In turn, reducing cognitive dissonance may improve health.

5.6. References

Abt, C. (1987). *Serious Games*. The Viking Press, New York, USA.

Alvarez, J., Djaouti, D., and Ramouroux, O. (2016). *Apprendre avec les Serious Games?* Editions Canopé, Poitiers, France.

Amato, É.A., (2007). Vers une instrumentalisation communicationnelle des jeux vidéo : quelles formes de séduction idéologique ou publicitaire ? *EUTIC 2007 International Conference Proceedings: "Enjeux et Usages des TIC"*, 7–10 November, Athens, Greece [Online]. Available at: http://www.omnsh.org/ressources/422/vers-une-instrumentalisation-communicationnelle-desjeux-video-quelles-formes-de.

Ashforth, B.E. and Tomiuk, M.A. (2003). *Emotional labour and authenticity: Views from service agents*. In *Emotions in Organizations*, Fineman, S. (ed.). Sage, London.

Frenay, M., and Bédard, D. (2011). Le transfert des apprentissages. *Apprendre et faire apprendre*. Presses Universitaires de France, Paris.

Gardner, R. (2013). Introduction to debriefing. *Seminars in Perinatology*, 37, 166–174.

Garris, R., Ahlers, R., and Driskell, J. (2002). Games, motivation, and learning: A research and practice model. *Simulation and Gaming*, 33(4), 441–467.

Heutte, J., Fenouillet, F., Boniwell, I., Martin-Krumm, C., and Csikszentmihalyi, M., (2014). Optimal learning experience in digital environments: Theoretical concepts measure and modelisation. *Symposium "Digital Learning in 21st Century Universities"*. Georgia Institute of Technology (Georgia Tech), Atlanta, Georgia.

Hochschild, A. (2003). Travail émotionnel, règles de sentiments et structure sociale. *Travailler*, 9(1), 19–49.

Hochschild, A. (2017). *Le prix des sentiments. Au coeur du travail émotionnel.* La Découverte, Paris.

Hou, H.T. and Chou, Y.-S. (2012). Exploring the technology acceptance and flow state of a chamber escape game–Escape The Lab© for learning electromagnet concept. *ICCE*, 38.

Juul, J. (2010). The game, the player, the world: Looking for a heart of gameness. *PLURAIS-Revista Multidisciplinar*, 1(2).

Leroux, Y. (2012). *Les jeux vidéo, ça ne rend pas idiot!* FYP Éditions, Limoges, France.

Lhuilier, D. (2006). Compétences émotionnelles : de la proscription à la prescription des émotions au travail. *Psychologie du travail et des organisations*, 12, 91–103.

Lhuilier, D. (2013). Introduction à la psychosociologie du travail. *Nouvelle revue de psychosociologie*, Eres, Toulouse, France.

Martin, L. (2017). Entraves à l'attitude ludique avec un jeu sérieux intégré dans une formation managériale : un exercice plus qu'un jeu ? *Sciences du jeu*, 7.

Martin, L. (2018). *Les Serious Games en Entreprise : Récréation ou Instrumentalisation Managériale*. Eres, Toulouse, France.

Martin, L. and Alvarez, J. (2018). Activité de serious gaming en formation de management d'équipe : réflexions autour des émotions suscitées et retours d'expériences. *Revue des interactions humaines médiatisées* 19(1), 29–46.

Mehl, J. (1981). Tricheurs et tricheries dans la France médiévale : l'exemple du jeu de dés. *Historical Reflections/Réflexions Historiques*, 8(2), 3–25. [Online]. Available at: http://www.jstor.org/stable/41298746.

Mayville, M.-L. (2011). Debriefing, the essential step in simulation. *Newborn and Infant Nursing Reviews*, 11(1), 35–39.

Michael, D.R. and Chen, S.L. (2005). *Serious Games: Games that Educate, Train, and Inform*. Muska and Lipman, Mason.

Morris, M.W. and Keltner, D. (2000). How emotions work: An analysis of the social functions of emotional expression in negotiations. *Review of Organizational Behavior*, 22, 1–50.

Pan, R., Lo, H., and Neustaedter, C. (2017). Collaboration, awareness, and communication in real-life escape rooms. *Proceedings of the 2017 Conference on Designing Interactive Systems*. ACM, 1353–1364.

Sanglerat, M.B., Grandjacques, B., and Francou, F. (2014). Les RPS et la méthode de la "situation-problème": l'organisation du travail à l'épreuve des risques psychosociaux, 111. [Online]. Available at: ///D:/Telechargement/12_rdctn1_mb.sanglerat-b.grandjacques-f.francou.pdf.

Sauvé, L., Renaud, L., and Gauvin, M. (2007). Une analyse des écrits sur les impacts du jeu sur l'apprentissage. *Revue des sciences de l'éducation*, 33(1), 89.

Schmidt, K. (1990). Analysis of cooperative work. A conceptual framework [Online]. Available at: https://papers.ssrn.com/sol3/papers.cfm?abstract_id=2015641.

Shouse, E. (2005). Feeling, emotion, affect. *M/C Journal*, 8(6) [Online]. Available at: http://journal.media-culture.org.au/0512/03-shouse.php.

Rebetez, C. and Betrancourt, M. (2007). Video game research in cognitive and educational sciences. *Cognition, Brain and Behavior*, 11(1), 131–142.

Taurisson, N. and Tchounikine, P. (2005). Une approche de l'apprentissage de l'organisation du travail collectif par la simulation, 153 [Online]. Available at: https://telearn.archives-ouvertes.fr/hal-00005657/document.

Turkle S. (1986). *Les Enfants de L'ordinateur*. Denoël, Paris.

Warmelink, H., Mayer, I., and Weber, J. (2017). Evaluating the team-building potential of a mixed reality escape room game. *Extended Abstracts Publication of the Annual Symposium on Computer-human Interaction in Play*. ACM, 111–123.

Welke, T., Leblanc, V.R., Savoldelli, G.L., Joo, H-S., Chandra, D.-B., Crabtree, N.A., and Naïk, V.N. (2009). Personalized oral debriefing versus standardized multimedia instruction after patient crisis simulation. *Economics, Education, and Policy*, 109(1), 183–189.

Zapf, D. (2002). Emotion work and psychological well-being – A review of the literature and some considerations. *Human Resource Management Review*, 12(2), 237–268.

6

The "Old" Issues of the "New" Artificial Intelligence Systems in Professional Activities

Based on scientific and "gray" literature as well as exploratory fieldwork, this chapter discusses several types of issues and questions raised by the "new" Artificial Intelligence (AI) systems in relation to work. After a discussion of what is covered by the term "AI", we describe the different functionalities and the envisaged or potential uses of these systems in work situations. Drawing on ergonomics literature, we show that these systems carry old issues that have been updated. Then, we present the questions of design and introduction of AI in work situations by underlining the important role of activity analysis to imagine and design human(s)–AI(s) systems in an organizational framework.

6.1. Introduction

For the last five years, the discourse on Artificial Intelligence (AI) has been omnipresent in the political and socio-economic fields. This phenomenon is notably linked to technological advances in AI in the field of machine learning, and has advanced due to the increase in machine power and the generation and availability of Big Data. According to many authors, the "new" AI systems are so powerful that it is now possible to automate an increasingly large number of tasks. Two perspectives therefore emerge: (1) the first is "optimistic" insofar as it emphasizes the (potential)

Chapter written by Tamari GAMKRELIDZE, Moustafa ZOUINAR and Flore BARCELLINI.

benefits that AI would bring to workers and companies. For example, it involves using AI to "free humans" from strenuous (repetitive or dangerous) tasks and/or to increase performance and productivity (Mcafee and Brynjolfsson 2017). (2) The second perspective is a product of fear, one that is mainly speculative: the massive diffusion of AI implies the replacement of humans in a large number of professional sectors (Frey and Osborne 2013)[1], or AI would be seen as a "super-intelligence" (Bostrom 2014) which will soon overtake humans in all areas where they exercise their intelligence.

Drawing on the scientific[2] and "gray"[3] literature as well as exploratory fieldwork[4], the objective of this chapter is to discuss several types of issues and questions raised by AI systems in relation to work. First, we clarify what is meant by the term AI. The next section presents the various functionalities and potential uses of "new" AI systems in work situations. Then, after having described the risks and promises generated by work-related AI, we show that these "new" systems carry old issues that have been updated. The final section focuses on issues of design and the introduction of AI systems in work situations. We will underline, on the one hand, the call for human-centered approaches to AI and, on the other hand, the important role of activity analysis and participatory project management in imagining, designing and deploying AI systems in an organizational framework.

6.2. AI: elements of definition and recent developments

Defining AI is not an easy task given the many different perspectives on it. On the one hand, AI can be a vague and abstract concept, even science fiction, and, on the other hand, a very concrete concept, referring to machines and algorithms with specific characteristics as well as a discipline at the crossroads of mathematics, computer science and cognitive science. The common point of the majority of

1 However, it is rather a question of transformations in professional activities (COE 2017) with the emergence of new professions (Wilson *et al.* 2017).

2 We mobilize work in ergonomics as well as in related disciplines.

3 Institutional and private practice reports, white papers, etc.

4 Participation in various events in 2019 on the theme of AI: for example, the BNP Paribas conference "*IA, nouvelles technologies, automatisation… Quelles incidences sur l'activité humaine ?*", the conference-debates of the research project "*Quelles régulations pour l'intelligence artificielle ?*", trade fairs and conferences such as NAIA, SIDO, Journées Francophones "Radiologue Augmenté", a workshop organized by Digital Society Forum entitled "*Le travail à l'ère de l'IA*" and exploratory talks with people directly or indirectly concerned by the emergence of AI.

definitions is the reference to humans, their intelligence and their cognitive abilities; even though the definition of human intelligence is the subject of debate (Legg and Hutter 2007) and the relevance of the term "Artificial Intelligence" is also questioned (De Ganay 2017). McCarthy (1985, p. 2), who is considered the "inventor" of the term "Artificial Intelligence" defines AI as "the science and engineering of making intelligent machines, specifically intelligent computer programs". The goal of this science is to create machines that are capable of simulating learning processes and human intelligence: using language, solving problems *a priori* reserved for humans and self-improvement through learning (McCarthy *et al.* 1955). To summarize, AI can be viewed from a number of different perspectives: it is both a discipline and a variety of computational techniques, based on the principle of "reproducing" human perceptual and cognitive abilities. In this chapter, the term "AI" will refer to a set of algorithms, machines and more broadly technologies in various forms (software, robotics, etc.) that are inspired by or aim to imitate human cognitive faculties such as the perception, production and understanding of natural language, the representation of knowledge, or reasoning.

Since the 1950s, two main currents have emerged: the symbolic approach and the connectionist approach (Mcafee and Brynjolfsson 2017). The symbolic approach, mobilized notably in the development of expert systems[5] from the 1960s, is based on the use of symbols and rules representing knowledge. The connectionist approach, dating from the 1940s, is inspired by neural networks; in this framework, the "representation" of knowledge and the general functioning of the system is based on nodes, coefficients and weights that are adjusted according to the type of task to be performed and the input data. It is this approach that is increasingly being developed today. While AI underwent significant development during the 1950s and 1960s, this period was followed by periods of a slowdown in research (known as the "winters" of AI) due mainly to the inability of systems to keep their promises. New development and new promises are now appearing, thanks to machine learning based on the connectionist approach and in particular one of its most popular branches today, called deep learning (Goodfellow *et al.* 2016). These techniques have been able to be reinvested in thanks to the increase in the computing power of machines and Big Data, essential for the operation of machine learning algorithms. Concretely, to perform a task, the machine "learns" from many examples prepared by a human (e.g. X-ray images to teach the system to recognize anomalies). This is called supervised learning and is the most commonly used techniques. Thus,

5 An expert system corresponds to "a software program based on knowledge derived from human expertise and capable on the one hand of proposing a solution to problems in a specific field and on the other hand of explaining the reasoning methods that led it to adopt this solution" (de Terssac *et al.* 1988, p. 463).

systems based on machine learning do not follow pre-set "rules" for performing operations. Using examples and specific algorithms, they develop "learning" models that are not explicitly programmed. As a result, these models evolve over time according to the training data and the data they process in an operational situation. Deep learning involves multi-layer artificial neural networks and enables computers to build complex concepts out of simpler ones (Goodfellow *et al.* 2016).

While these AI systems have made significant progress in many areas (e.g. computer vision or natural language processing), they have two significant limitations: (1) a lack of explainability, which describes the difficulty in understanding how learning-based systems work when they produce a specific output, especially if based on deep learning. Specifically, experts in the field are still unable to precisely explain what happens when such systems "learn" (Zouinar 2020). Research is underway to make these systems explainable, that is, systems whose outcomes can be "explained". The question of explainability was already raised with older generation AI systems (in particular, expert systems) but in a less difficult way. (2) The use of the concept of learning is, at best, a metaphor. A learning system does not learn in the same sense that humans can. It cannot take initiative and is not autonomous outside a framework predefined by a series of programs and instructions that mark out the range of possibilities. We therefore speak of "weak" AI, specialized in specific tasks, as opposed to "strong" AI, which refers to autonomous and generalized "intelligent" systems with "common sense" and "consciousness" (CNIL 2017). Currently, however, "strong" AI does not exist.

6.3. Functionalities and (potential)[6] uses of new generation AI systems

The purpose of this section is to clarify what "new" AI systems are – or can be – from the point of view of both their functionality and their use. Functionality here means what these systems do by design. We present an overview (synthetic and non-exhaustive) of systems based on machine learning algorithms. A lot of professional sectors are concerned by the diffusion of AI systems: industry, defense and security, justice, education, health, customer relations, finance, transport, environment, agriculture, human resources, etc. In some areas, these systems are already deployed, in others – which constitute the majority – they are still in the research phase. The functionalities of these AI systems can be presented according to four types (see Figure 6.1).

6 *A priori*, the systems described here exist but we do not have precise and empirical elements concerning their use in real situations.

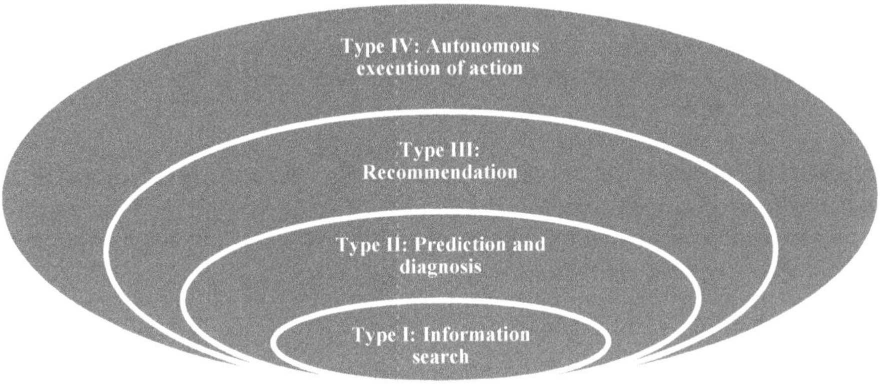

Figure 6.1. *Types of functionality of "new" generation AI systems*

The first type of AI system is limited to tools for information retrieval in massive databases. The second type includes more sophisticated systems capable of performing calculations of the probabilities of occurrence of events and/or of making a diagnosis. Systems of the third type go beyond prediction. Based on their analyses and predictions, they can propose possible courses of action to the user or recommend "decisions". Finally, the fourth type includes systems that are capable of making a decision or acting partly autonomously without human intervention. Let us illustrate these types of functionalities with real-world examples of existing or imagined applications in different professional sectors.

In the banking and insurance customer relationship, Crédit Mutuel's[7] account managers[8] have a support tool based on the IBM's[9] AI Watson system. The aim is to help them search for information in their internal document base via a dialogue in "natural" language. Here, we have an AI system with the first type of functionality, which is limited to information retrieval to provide an answer to a question asked. The system functions like a sophisticated search engine which is supposed to facilitate the account managers' work. The second type of functionality can be illustrated by predictive/preventive maintenance systems. This involves calculating the risk of future malfunctions or estimating the elements and parts of machines that will require maintenance in the near future. Several systems of this type exist on the market today, deployed in various fields such as that of transport or industry. For

7 A French bank.

8 https://www.creditmutuel.com/fr/actualites/watson-nouvelle-solution-innovante-au-service-du-reseau-et-de-la-relation-client.

9 Watson is an IBM ecosystem consisting of several tools, algorithms and specialties that can be used as a basis for the development of AI systems. See: https://www.ibm.com/watson.

example, in the field of motorization, Otosense[10] developed a system that enables preventive maintenance based on engine noise analysis. The system is supposed to reduce machine downtime and maintenance costs, and avoid accidents. In the legal field, the Predictice[11] system makes it possible to estimate the success rate of a legal action and the associated indemnities. Predictive systems are also offered in the field of agriculture. For example, Microsoft, in partnership with The Yield[12], offers an AI tool that gives farmers accurate micro-local weather forecasts based on a dozen factors (soil moisture, wind, rain, etc.) to help them make decisions (e.g. planting decisions). Concerning the third type of functionality, many AI systems that deliver this functionality have been developed for the health field, such as medical research, psychiatry or medical imaging. Regarding medical imaging, this development is due in particular to the increased performance in image recognition and analysis. For example, Arterys[13] develop systems with several specialized algorithms to automatically analyze radiological images, perform measurements on them or detect variations and anomalies. Thus, the system can, for example, automatically detect fractures and can recommend to doctors which emergency X-ray images to look and interpret first. Finally, there are also AI-based support tools with more advanced execution order functionalities, that is, they can partly act "autonomously". This is the case of the JulieDesk virtual assistant[14], which is able to manage a schedule such as sending e-mails to propose available slots and thus organize a meeting.

It can therefore be seen that a wide variety of AI applications in different professional sectors is offered, even if the majority of systems we have identified remain at the prototype and use scenario stage, or in the research phase. While much of the discourse on these systems focuses on the expected benefits, for example, in terms of economic performance and in some cases improvement of the quality of working life, AI raises a number of questions. In addition to ethical and legal issues[15] as well as the utility and expected benefits, other questions arise in relation to the consequences of the introduction of these systems on work and its organization, the transformation of the professions concerned, the reconfiguration of work activities and organizations. Another set of questions concerns the place that these systems should occupy in human activities: how should we think about the distribution of roles or tasks between these systems and humans? What is the place and the share of autonomy of workers in their activities when they are confronted with systems that are potentially more efficient than themselves, or which can guide

10 https://www.otosense.com/.

11 https://predictice.com/.

12 https://news.microsoft.com/transform/videos/yield-feed-world-without-wrecking-planet/.

13 https://www.arterys.com/.

14 https://www.juliedesk.com/fr/.

15 Due to lack of space, a discussion of ethical and legal issues cannot be explored here. For a more detailed overview, see the CNIL report (2017).

the workers in making irrelevant decisions? What type of skills must be developed to work with these machines? What about uses in real work situations? Finally, there are two other important interrelated issues that have to do with the introduction of these systems in work situations: the design of these systems and the management of the transformation of work activities that may be introduced. In the rest of this chapter, we address some of these questions, drawing on previous ergonomics literature related to AI and automation.

6.4. The "new" generation of AI and the old challenges of transforming work situations

The "new" AI systems are seen as a technological evolution that will lead to a new wave of automation[16] since it is assumed that it will affect a very wide range of professions. In order to discuss the current questions and issues raised by AI systems, we mobilize studies conducted on old AI systems, as well as on automation whose consequences on human activity at work are well documented. These studies have led to the confrontation of two approaches that we will briefly develop below[17]: the substitution of humans by machine, and human–machine "complementarity".

6.4.1. *From the substitutive approach of automation to the complementary approaches of human–machine systems*

The idea of substitution is to replace a human with a machine in some of their tasks, where the machine is supposed to be more efficient and less expensive than workers. One popular approach to automation consists of distributing tasks between humans and machines on the basis of a comparison and opposition of their strengths and weaknesses. These models, of the MABA–MABA (Men Are Better At–Machines Are Better At) type, have been the subject of much criticism, mainly because they are static and too generalized (Dekker and Woods 2003) and because they are generally based on a task breakdown that does not take into account actual work (Bainbridge 1983). Broadly speaking, several drawbacks of the introduction of automated systems in the workplace, based on these models, have been identified. For example, when they place humans outside the decision-making loop, humans are placed in a passive position with many consequences, ranging from decreased alertness, loss of skills and a loss of situation awareness (Sarter *et al.* 1997;

16 Automation can be defined as "the execution by a machine agent (usually a computer) of a function that was previously carried out by a human" (Parasuraman and Riley 1997, p. 231).

17 For more details on these two approaches, see Zouinar (2020).

Parasuraman and Wickens 2008). The opacity[18] of automated systems is another problematic element affecting the understanding and resolution of the problem by humans (Roth *et al.* 1987), the trust given to these systems, their uses and also their recovery in case of problems (Parasuraman and Riley 1997). Issues at the organizational and collective level have also been raised, for example, in the case of expert systems (de Terssac *et al.* 1988). By formalizing actual practices of work, these systems were supposed to lead workers to lose control over the actual organization of their work.

According to the complementarity approaches, the machine is supposed to assist the human in their work, that is, the machine is not meant to replace them but rather in some cases it is meant to carry out some of their tasks. This could be seen as a partial substitution, but the important point here is that the aim of the system is to *help* the human. Another important point is that, in this approach, these "high-level intelligent automatisms" are seen as "teammates" rather than tools (Amalberti 2002). Another complementary approach is that of symbiosis (Licklider 1960). This notion implies a "co-evolution", a "co-adaptation" or even a "co-development" of two entities – human and machine – that influence each other. In moving towards these perspectives, different models emerged in the 1980s. Instead of taking the human out of the decision-making loop, automation models were refocused on the human and their activity (Billings 1996). Other models assume that humans and machines could have a "cooperative" or "collaborative" relationship (Hoc 2003; Woods *et al.* 1990). However, this would require automated systems to be equipped with communication and cooperation capabilities, such as understanding the intentions, goals and actions of human agents. As systems do not today have these complex capabilities that are characteristic of human cooperation[19], the use of the term "cooperation" or "collaboration" is simply metaphorical. Moreover, these approaches do not take into account the asymmetry between human and artificial agents (Suchman 1987; Salembier and Pavard 2004). For example, unlike humans, a machine has, at best, very limited access to the relevant context of action. Finally, another criticism of these approaches is their tendency to focus on the activity of a single human agent in their relationship with an artificial agent. They neglect the collective and organizational dimensions of the activity in human–machine systems.

In the following section, in light of this background information, we will look at the many issues and questions generated by the "new" AI systems.

18 Introduced by Billings (1996), the term "opacity" refers to a lack of capacity of systems to transmit to humans clear and understandable information and explanations about how they function and the processes they carry out.

19 For more on this point, see Barcellini (2020) and Zouinar (2020).

6.4.2. *The challenges of "new" AI systems in work situations*

First of all, AI raises questions about new forms of organization. The introduction of these systems would not simply be a technological tool of improving work processes but rather a way of rethinking the organization itself. These future organizations are imagined to be more transparent, less hierarchical and more cooperative thanks to AI systems which are supposed to be "a vertical sharing factor, giving each level visibility on all the activities of the different levels of the organization, and horizontal by aggregating data from the different entities of the organization" (Giblas *et al.* 2018, p. 40). According to the authors, it would bring about new perspectives of power and information sharing.

Secondly, there is the issue that has to do with the place of AI in work activities. Different approaches have been proposed. For example, instead of considering machines as providers of information to humans to help them make decisions, an alternative approach proposes that it is the human who could provide judgment and advice to machines making decisions (Mcafee and Brynjolfsson 2017). All automatable tasks would be delegated to machines with the human out of the decision-making loop, except in new and unusual situations where the human would have the opportunity to question the machine's decisions. Another approach suggests that, for problem-solving tasks, it would make more sense to systematically mobilize AI systems first, giving humans a supervisory role: "where [teams] were in the execution of procedures, they are now in charge of supervising the work of machines, ensuring their continuous learning" (Tata Consultancy Services 2018, p. 15). This view of the division of roles between humans and machines is not new and raises many problems. By reducing the worker's activity, these models once again place the human in a passive position, as it was already the case with the early automation models. As we saw in section 6.4.1, this passive position is a source of problems, such as loss of skills, autonomy and an overall "situation awareness" (Endsley and Kiris 1995). Thus, from the perspective of activity-centered ergonomics (Guérin *et al.* 2007), the objective would be to ask how the "new" AI systems can constitute resources for the work activity, how they can be integrated into the activity and the organization and how humans can work efficiently and healthily with these systems. Thought and attention must also be given both to the risks of obsolescence of the human's skills, whose activity is transformed by the arrival of these systems, and to the development of new skills needed to use AI systems (Mateescu and Elish 2018). This human-centered vision, as we have seen, leads to human–machine complementarity approaches that have already been thoroughly discussed. These approaches consider the human–AI relationship in terms of "cooperation", "collaboration", "augmentation" and "symbiosis". The augmentative perspective – initiated by Engelbart (1962) with the idea of human intelligence augmented by machines – is notably present in Daugherty and Wilson's (2018) model. The authors imagine the existence of a "missing middle" between non-automatable occupations requiring creativity, leadership and empathy, and occupations characterized by repetitive and routine tasks that can potentially be automated. In this missing middle, there are "hybrid" activities carried out by humans and machines that augment each other. This idea of augmentation is taken up by others such as Jarrahi (2018) who advocates a human–AI "symbiotic" relationship in organizational decision-making. In these perspectives, AI is seen as a tool for augmentation rather than automation.

The development of human skills is also often approached from this angle of complementarity with the capabilities of machines, in the sense that it is up to humans to adapt and develop skills that complement AI, in other words skills that machines do not yet have (Villani 2018). This approach is problematic if the complementarity is mainly focused around the abilities of the system, the human seen as an "auxiliary part" that has to adapt themselves to the machine's abilities. Some argue that the time (supposedly) freed up by the automation of certain "routine" tasks should be devoted to the development of "soft skills" (Lemberger *et al.* 2019) such as social and relational skills as well as creative and precision skills. In this way, humans would become complementary to AI and, in turn, AI would increase its capabilities. This idea of augmenting humans through AI is widely mobilized in the discourses of companies and consulting firms as it implies an increase in human performance.

The majority of models of the complementarity of humans and AI, developed in recent years, make very little reference to previous work on human–machine "cooperation" or "collaboration". Moreover, despite the advances of AI systems in terms of performance, the limits of the approaches to human–machine "cooperation" or "collaboration" remain valid (Van Den Bosch and Bronkhorst 2018; Barcellini 2020; Zouinar 2020)[20]. For example, the limited capacity of AI to understand the context of activity still does not allow us to really talk about a collaborative relationship between humans and AI. Mutual understanding of context is a necessary element for a collaborative relationship: AI must be able to identify the relevant characteristics of the human partner and the situation, understand their goals and actions; conversely, the human may need to understand how AI works and the results it provides, which seems to be difficult in the current context where AI systems are characterized by a lack of explainability. Without these conditions, the interaction between humans and AI is limited to unidirectional interactions (e.g. when a system gives a result to humans without explanation and without taking into account their goals) and is therefore not really a collaboration. True collaboration between these two entities as members of a team could be possible if both members have the ability to share at least common goals, mutual adaptation, common

20 To our knowledge, they are the only current authors referring to the work of the 1980–1990s on human–machine and expert systems.

understanding of the situation or interdependent relationships (Salembier and Pavard 2004; Johnson *et al.* 2014; Zouinar 2020). Finally, speaking of human–AI "cooperation" or "collaboration", which for the moment proves to be unattainable, carries risks, in the sense that it may violate the expectations of the human partner and may hinder their activity.

To conclude, these complementarity approaches have many limitations, but they have the merit of focusing on the human and rejecting the purely substitutive approach. Given the state of the art in AI, a minimal conclusion is that it remains important to develop AI systems that assist workers in their activities; that these systems would be understandable to them for all practical purposes, and that workers leverage their skills, abilities and performance. While the question of defining the role of AI in relation to the human and vice versa is central in a complementarity perspective, it cannot be dealt with independently of the design process of these systems or of the modalities of their integration in work situations. It is this point that we will now discuss.

6.5. What are the approaches to designing and integrating AI systems in work situations?

AI design and deployment approaches are mostly part of a "techno-centric" approach which consists of focusing on technology as a solution and dealing with all the issues concerning humans and their work in the background. This approach has been widely criticized for the residual position it gives to the human and their actual work vis-à-vis the technical tool (Rabardel 1995; Woods 1996). However, some "projective" approaches claim to distance themselves from this approach by emphasizing the need to give an important place to workers in the design and integration of AI in the workplace. By way of illustration, let us examine some of these approaches.

The approach proposed by Nerce *et al.* (2018) for the adoption of AI in organizations emphasizes the objective to "keep humans at the heart of transformation" (*ibid.*, p. 82). This approach claims to respond to the human issues through an acculturation program with the aim of demystifying AI and engaging employees in the transformation project, or through the creation of "AI labs" to identify and create AI use cases and the transmission of knowledge about these systems and their uses. These types of approaches, stemmed from projective and hypothetic reflections, show a willingness to demonstrate the importance of issues relating to the consequences of the introduction and use of AI on humans, their work and organization. However, the actual consideration of these issues is very limited or remains wishful thinking, and is strongly oriented by the technological vision of the projects. Indeed, most of these projective reflections often consider AI projects as technological projects that do not really address organizational issues or those related to the actual activity. However, a technological change is never a simple change of tool. It leads to profound transformations in workplace activities and organizations that should be taken into account. As shown by a long tradition of ergonomics research and interventions, the analysis of effective work is particularly useful and even necessary to take these kinds of transformations into account. For example, going in that direction and drawing on a socio-technical approach, Willis and Jarrahi (2019) mobilized the analysis of work to propose a division of labor between a general practitioner and an AI system for the processing of medical records. This analysis makes it possible, in particular, to understand and determine the aspects of the doctor's work that can be automated so that AI does not prevent the activity from running but becomes a resource for the practitioner.

Concerning the design of AI systems, Xu (2019) proposes to develop human-centered AI (HAI)[21]. The basic principles of this approach follow three objectives: to technically "reflect" the characteristics of human intelligence, to support rather than replace the human and to pay particular attention to the impacts of AI on humans (designing AI systems that are explainable, useful and usable). Inspired by approaches to human–machine interaction (notably, user-centered design), the design of AI systems according to this approach is mainly considered with its criteria of explainability, usefulness and usability, without, however, taking into account the collective and organizational dimensions of work that will inevitably be transformed.

Generally speaking, we consider that the ergonomic design approach (Barcellini *et al.* 2014) grounded in the analysis of the work activity and the simulation of future activity (Béguin and Cerf 2004) can constitute a real contribution to meet these challenges of work activity transformation. With a participative and collaborative approach (involving all stakeholders and creating the conditions for their collaboration), this ergonomic approach consists of building on what already exists, that is, analyzing the context, the environment and the work activity in a systemic approach (Dul *et al.* 2012). With the objective of building a framework for projecting the situation of the future, this analysis allows the identification of work situations that will be mobilized in simulation aimed at exploring the design space as well as the potential transformations of the work activity. More concretely, these simulation situations consist of making the relevant actors "play" their own activity in a new context that includes AI systems. Thus, this approach makes it possible to imagine the technical design of systems upstream, and the collective and organizational transformations it implies.

---

21 HAI is also a research institute recently established by Stanford University, Berkley and MIT.

6.6. Conclusion

The unprecedented performance and scope of the "new" AI systems in a growing number of professional activity sectors have led us to address the issues raised by the introduction of these technologies in work situations. We have seen that the majority of these issues are old and have been studied many times in ergonomics and other disciplines concerned with work. However, due to the constant improvement of the performance of current AI systems based on machine learning, some of these issues have been brought up to date in a more significant way, such as the division of labor between humans and machines that are able to "learn", or the problem of explainability.

At present, these issues are being addressed more from an economic or technical point of view, whether by private companies or public institutions. Their discourse is often aimed at promoting these systems, their advantages in terms of performance and productivity gains, or the reduction of arduousness and routine tasks "that don't have added value" for employees. In contrast to the substitutive approaches of automation, some of these discourses also increasingly put forward the augmentative human approach through technology. The terms "augmented employee", "augmented manager" and "augmented doctor" are emerging. This idea of augmentation is interesting as it is focused on the benefits of technology for workers, but it is not always clear as to what is really augmented. It might be more accurate to think of the empowerment of workers as an "augmentation" in the sense of supporting human abilities to perform existing or new work activities. The terms of "cooperation", "collaboration" and "symbiosis" used alongside AI are also strongly mobilized. However, we have seen that AI systems do not yet have real collaboration or cooperative capacities, and even less characteristics such as autonomy to be in symbiosis with humans. While the idea of seeing these systems as "partners" is quite appealing (Norman 2017), it must not, however, lead to the symmetrization of the human and the machine, that is, we must not consider them as equal partners. Finally, one of the central issues that is not often addressed is the approach to designing and integrating AI systems in work and organizations. To take advantage of AI technologies when they are potentially useful in a given situation, the ergonomic approach grounded in real activity analysis and the simulation of future activity seems to us to be a particularly relevant and promising approach for developing AI systems that are conducive to human activities and that are better integrated into organizations.

6.7. References

Amalberti, R. (2002). La conception d'automatismes et les systèmes tolérants aux erreurs Automation Design and Error Tolerant Systems. *PRTO HFM Specialists' Meeting on "Human Factors in the 21st Century"*, Paris.

Bainbridge, L. (1983). Ironies of automation. *Automatica*, 19(6), 775–779.

Barcellini, F. (2020). Quelles conceptions de la coopération humains – robots collaboratifs ? *Activités*, 17(1).

Barcellini, F., Van Belleghem, L., and Daniellou, F. (2014). Design projects as opportunities for the development of activites. In *Constructive Ergonomics*, Falzon P. (ed.). Taylor and Francis, New York, USA.

Béguin, P. and Cerf, M. (2004). Formes et enjeux de l'analyse de l'activité pour la conception des systèmes de travail. *Activités*, 01(1), 54–71.

Billings, C.E. (1996). *Human-centered Aviation Automation: Principles and Guidelines.* Ames Research Center, California, USA.

Bostrom, N. (2014). *Superintelligence*. Oxford University Press, Oxford.

CNIL (2017). Comment permettre à l'Homme de garder la main ? Report, CNIL, Paris.

COE (2017). Automatisation, numerisation et emploi. Tome 1 : les impacts sur le volume, la structure et la localisation de l'emploi. Report, COE, Paris.

Daugherty, P.R. and Wilson, J. (2018). *Humans + Machine: Reimagining Work in the Age of AI.* Harvard Business Review Press, Boston, USA.

Dekker, S.-W.-A. and Woods, D.D. (2003). MABA-MABA or abracadabra? Progress on human-automation co-ordination. *Cognition, Technology and Work*, 4(4), 240–244.

Dul, J., Bruder, R., Buckle, P., Carayon, P., Falzon, P., Marras, W. S., and van der Doelen, B. (2012). A strategy for human factors/ergonomics: Developing the discipline and profession. *Ergonomics*, 55(4), 377–395.

Endsley, M.R., and Kiris, E.O. (1995). The out-of-the-loop performance problem and level of control in automation. *Human Factors: The Journal of the Human Factors and Ergonomics Society*, 37(2), 381–394.

Engelbart, D.C. (1962). Augmenting human intellect: A conceptual framework. SRI Summary Report AFOSR-322 Air Force Office of Scientific Research, Washington, D.C., USA.

Frey, C.B. and Osborne, M.A. (2013). *The Future of Employment: How Susceptible are Jobs to Computerisation?* University of Oxford, Oxford.

de Ganay, C. (2017). Pour une intelligence artificielle maîtrisée, utile et démystifiée. L'office parlementaire d'évaluation des choix scientifiques et technologiques. Report, Sénat, Paris.

Giblas, D., Godon, A.-S., Fargeas, M., Duranton, S., Gard, J.-C., Audier, A., Caye, J-M., and Buffard, P.-E. (2018). *Intelligence artificielle et capital humain. Quels defis pour les entreprises ?* Malakoff Médéric and Boston Consulting Group, Boston, USA.

Goodfellow, I., Bengio, Y., and Courville, A. (2016). *Deep Learning*. The MIT Press, Massachusetts, USA.

Guérin, F., Laville, A., Daniellou, F., Duraffourg, J., and Kerguelen, A. (2007). *Understanding and Transforming Work: The Practice ef Ergonomics*. ANACT Network Edition, Lyon.

Hoc, J.-M. (2003). Coopération humaine et systèmes coopératifs. In *Ingénierie cognitive. IHM et cognition*, Boy G. (ed.). Hermès-Lavoisier, Paris.

Jarrahi, M.H. (2018). Artificial intelligence and the future of work: Human-AI symbiosis in organizational decision-making. *Business Horizons*, 61(4), 577–586.

Johnson, M., Bradshaw, J.M., Feltovich, P.J., Jonker, C.M., Van Riemsdijk, M.B., and Sierhuis, M. (2014). Coactive design: Designing support for interdependence in joint activity. *Journal of Human-Robot Interaction*, 3(1), 43–69.

Legg, S. and Hutter, M. (2007). Universal intelligence: A definition of machine intelligence. *Minds and Machines*, 17(4), 391–444.

Lemberger, P., Parent, N., Scialom, T., and Mauléon, F. (2019). *Vers une nouvelle alliance homme-machine ?* onepoint x weave [Online]. Available at: https://weave.eu/vers-une-nouvelle-alliance-homme-machine/.

Licklider, J.C.R. (1960). Man-computer symbiosis. *IRE Transactions on Human Factors in Electronics, HFE-1*, 4–11.

Mateescu, A. and Elish, M.C. (2018). *AI in Context: The Labor of Integrating New Technologies*. Data and Society [Online]. Available at: https://datasociety.net/wp-content/uploads/2019/01/DataandSociety_AIinContext.pdf.

McAfee, A. and Brynjolfsson, E. (2017). *Machine Platform Crowd: Harnessing our Digital Future*. W.W. Norton and Company, New York, USA.

McCarthy, J. (1985). What is artificial intelligence? *American Scientist*, New Haven, CT, USA.

McCarthy, J., Minsky, M.L., Rochester, N., and Shannon, C.-E. (1955). A proposal for the Dartmouth Summer Research Project on artificial intelligence, August 31, 1955. *AI Magazine*, 27(4), 12.

Nerce, J., Marie, S., and Coggia, P. (2018). (Re)launch, scale and lean AI with cloud. White paper, Artefact, Paris.

Norman, D. (2017). Design, business models, and human-technology teamwork. *Research-Technology Management*, 60(1), 26–30.

Parasuraman, R. and Riley, V. (1997). Humans and automation: Use, misuse, disuse, abuse. *Human Factors*, 39(2), 230–253.

Parasuraman, R. and Wickens, C.D. (2008). Humans: Still vital after all these years of automation. *Human Factors*, 50(3), 511–520.

Rabardel, P. (1995). *Les Hommes et les technologies ; approche cognitive des instruments contemporains*. Armand Colin, Paris.

Roth, E.M., Bennett, K.B., and Woods, D.D. (1987). Human interaction with an "intelligent" machine. *International Journal of Man-Machine Studies*, 27(5–6), 479–525.

Salembier, P. and Pavard, B. (2004). Analyse et modélisation des activités coopératives situées. *Activités*, 1(1), 87–99.

Sarter, N.B., Woods, D.D., and Billings, C.E. (1997). Automation surprises. *Handbook of Human Factors and Ergonomics*, 29–31.

Suchman, L.A. (1987). *Plans and Situated Actions: The Problem of Human–Machine Communication.* Cambridge University Press, Cambridge.

Tata Consultancy Services (2018). Les entreprises françaises sur la voie de l'intelligence artificielle. Study, Tata Consultancy Services, Paris.

de Terssac, G., Soubie, J., and Neveu, J. (1988). Systèmes experts et transferts d'expertise. *Sociologie du travail*, 3, 461–477.

Van Den Bosch, K. and Bronkhorst, A. (2018). Human-AI cooperation to benefit military decision making. *STO*, 1–12.

Villani, C. (2018). Donner un sens à l'intelligence artificielle [Online]. Available at: https://www.aiforhumanity.fr/pdfs/9782111457089_Rapport_Villani_accessible.pdf.

Willis, M. and Jarrahi, M.H. (2018). Automating documentation: A critical perspective into the role of artificial intelligence in clinical documentation. iConference, Washington, D.C., USA.

Wilson, H.J., Daugherty, P.R., and Morini-Bianzino, N. (2017). The jobs that artificial intelligence will create. *Mit Sloan Management Review*, 58(4), 14–16.

Woods, D.D. (1996). Decomposing automation: Apparent simplicity, real complexity. In *Automation and Human Performance: Theory and Applications*, Parasuraman R. and Mouloua M. (eds), 3–17.

Woods, D.D., Roth, E.M., and Bennett, K.B. (1990). Explorations in joint human–machine cognitive systems. *Human–Machine Cognitive Systems*, 123–158.

Xu, W. (2019). Toward human-centered AI: A perspective from human–computer interaction. *Interactions*, 26(4), 42–46.

Zouinar, M. (2020). Evolutions de l'intelligence artificielle et travail : quels enjeux pour l'activité humaine et la relation Humain-Machine ? *Activités*, 17(1).

PART 2

New Modalities and Forms of Work…

7

Challenges in Deploying Telework: Benefits and Risks for Employees

Based on research into the impacts telework has on employees, this chapter aims to report on the challenges of deploying telework within work organizations. After defining telework and the situations it covers, we will discuss both the benefits and the constraints associated with this mode of work. Finally, in view of the difficulties and risks identified in the scientific literature in this field, we will report on the stages that are likely to mark out the approaches to the deployment of telework in work structures.

7.1. Telework: definitions and characteristics

To our knowledge, there is no consensus in the literature in the field on how to define telework. Some authors adopt a very broad definition and consider all remote work and mediated work situations as telework situations, whether formal or informal (Taskin and Devos 2005; Troup and Rose 2012), cyclical, regular or permanent (Taskin and Bridoux 2010; Coenen and Kok 2014). However, some studies point out that occasional or informal telework is more often associated with overtime work or the need to catch up on a file or project (Ojala 2011). Using this definition, it would therefore be more characteristic of an extension of work that is made possible by the use of digital technologies. In addition, empirical studies that seek to assess the effects of telework focus more specifically on regular telework, which is also more likely to be formalized or contractualized within the work organization (Troup and Rose 2012).

Chapter written by Emilie VAYRE.

With regard to the criteria that are most frequently proposed in the literature (i.e. characterization of telework situations and criteria for inclusion of survey respondents), telework can be considered as a generic term that refers to a regular and formal form of organizing and carrying out professional activities that is carried out totally or partially remotely (i.e. carried out outside the employer's premises at least one day a week) and through the use of information and communication technologies (Vayre 2019). Far from representing a homogeneous work arrangement, telework can be characterized according to four broad categories (e.g. Taskin and Devos 2005; Hislop and Axtell 2007; Mello 2007; Vayre 2019):

– full-time or permanent (i.e. carried out exclusively or almost exclusively at home) home-based teleworking;

– teleworking at home on an alternating, commuting or part-time basis (i.e. at home at least one day a week);

– nomadic or mobile teleworking (i.e. teleworking during business trips and combining different locations: hotel, home, clients' premises, transport, etc.);

– finally, teleworking in dedicated third locations, that is, in tele-centers, satellite offices or coworking spaces (i.e. carried out in premises dedicated to work, located away from the main company location and, in principle, close to the employee's place of residence).

It should be noted that the most common form of telework, whether permanent or commuting, is home-based (Worldatwork 2011; Bentley *et al.* 2016; Bureau of Labor Statistics 2016).

7.2. The benefits of teleworking

Many studies highlight the positive impact of telework on performance and the conditions for carrying out the work (provided that it is not associated with a strengthening of managers' control over the activity; Solís 2017). Indeed, telework makes it possible, on the one hand, to reduce noise pollution, interruptions and distractions specific to the "usual" working environment in organizations, and, on the other hand, to increase concentration, professional efficiency, quality of work, performance and productivity (e.g. Montreuil and Lippel 2003; Mello 2007; Martin and MacDonnell 2012; McNaughton *et al.* 2014; Vega *et al.* 2015; Biron and van Veldhoven 2016).

Telework is also associated with lower absenteeism, lower levels of intent to leave the company and lower staff turnover rates (e.g. Gajendra and Harrison 2007; Mello 2007; Martin and MacDonnell 2012). Moreover, it allows employees to better cope with and regulate job stress, whether it is generated by high work demands (overwork, role conflict) or a deleterious work environment (Duxbury and Halinski 2014; Biron and van Veldhoven 2016).

Additional results show that telework promotes a sense of control over one's working time, perceived flexibility and a level of control over how the task is performed or work is organized and carried out on a daily basis (e.g. Maruyama *et al.* 2009; Sardeshmukh *et al.* 2012; McNaughton *et al.* 2014). It promotes a sense of autonomy, work motivation, organizational involvement and job satisfaction, especially when teleworkers have support from their supervisors and organizational support (e.g. Gajendra and Harrison 2007; Kelliher and Anderson 2010; Martin and MacDonnell 2012; Troup and Rose 2012; Vega *et al.* 2015).

Finally, telework is also associated with a better balance between work and "not work" time (Gajendra and Harrison 2007). Indeed, it allows better management, organization and prioritization of the various activities to be carried out in different areas of life, better reconciliation of the multiple roles to be assumed and, ultimately, greater presence in the family sphere (e.g. Metzger and Cléach 2004; Maruyama *et al.* 2009; Sardeshmukh *et al.* 2012; Duxbury and Halinski 2014; Vayre and Pignault 2014).

7.3. The constraints and risks of teleworking

For many authors, the results observed in studies on the gains in productivity, efficiency and quality of work among teleworkers are, in fact, due to an intensification, densification and extension of work (Montreuil and Lippel 2003; Metzger and Cléach 2004; Taskin and Devos 2005; Maruyama *et al.* 2009; Sardeshmukh *et al.* 2012). Kelliher and Anderson (2010) argue that the increase in working time associated with telework can be explained by social exchange theory: in many companies, telework is considered a privilege and is not available to everyone, and teleworkers would feel indebted to their organization and would make more effort to pay their debt. Metzger and Cléach (2004, p. 443), for their part, show that under the pressure of "internalized guilt" (guilt about being able to work without the constraints of transport, noise pollution, etc.), the teleworker is compelled to work without limits. The reduction in interruptions and breaks observed among teleworkers leads them to be exposed to certain occupational risks, such as musculoskeletal disorders, for longer periods (Montreuil and Lippel 2003). Telework is thus recognized as a source of overwork, "workaholism", occupational stress and even burnout (Montreuil and Lippel 2003; Metzger and Cléach 2004; Peters *et al.* 2008).

Teleworkers' time investment in their work also creates difficulties in limiting it and not letting it overwhelm them (e.g. Metzger and Cléach 2004; Taskin and Devos 2005; Wilton *et al.* 2011; McNaughton *et al.* 2014). Telework also involves a spatio-temporal intertwining of professional and personal activities, which disrupts their fulfillment and blurs the boundary between these two spheres of existence (Vayre and Pignault 2014). The emergence of misunderstandings, tensions and conflicts with family and friends, as well as increased stress in the private sphere, are also frequently observed (e.g. Metzger and Cléach 2004; Taskin and Devos 2005; Wilton *et al.* 2011; McNaughton *et al.* 2014).

In terms of relationships with their professional environment, empirical research shows that teleworkers have less frequent, fewer and lower quality exchanges and discussions with colleagues and supervisors when they are away from their workplace (Gajendra and Harrison 2007; McNaughton *et al.* 2014; Vayre and Pignault 2014). Teleworking is thus said to impair communication and cooperation within work teams, which may also lead to feelings of isolation, uncertainty, stress and even burnout (Montreuil and Lippel 2003; Bentley *et al.* 2016). The risk of professional and social isolation as well as the fragmentation of work collectives is thus often mentioned (Taskin and Devos 2005; Hislop and Axtell 2007; Wilton *et al.* 2011). Teleworkers report a lack of informal or spontaneous interactions with colleagues and a feeling of exclusion from the work organization associated with a feeling of loneliness and professional isolation (e.g. Cooper and Kurland 2002; Golden *et al.* 2008; McNaughton *et al.* 2014).

Greer and Payne (2014) also find that, in the face of social pressure, teleworkers develop behaviors to remain visible to their professional entourage in order to prove that they are working. Indeed, social representations concerning telework are often negatively connoted: misunderstanding, mistrust or even hostility may be reflected in comments from professional circles, which no longer have the usual visual indicators showing the professional investment and activities actually carried out by the teleworker (e.g. Greer and Payne 2014; Vayre and Pignault 2014).

As a result of this physical separation, teleworkers are also less exposed to the norms, rules and values of the company, and, as a result, they are less attached, less committed, more independent and have less sense of belonging and less identification with the work organization (Hislop and Axtell 2007; Mello 2007; Golden 2009; Sardeshmukh *et al.* 2012). Finally, according to Taskin and Bridoux (2010), the development of telework also affects organizational socialization, as exchanges between teleworkers and non-teleworkers are less frequent and mediatized and remote communication makes it more difficult to transmit and share information, knowledge and skills. Several studies also mention the influence of this geographical and socio-professional distance on the career development of teleworkers, which tends to stagnate as opportunities for advancement and promotion are less accessible due to their physical absence (Cooper and Kurland 2002; Gálvez *et al.* 2012).

7.4. The challenges of deploying telework in organizations

As mentioned above, telework is a major issue when considering its impact on employees. Therefore, before and during its deployment, it is important to consider the opportunities and risks of telework:

– work organization, professional involvement and efficiency (i.e. spatio-temporal management of individual and collective activities, identification with company standards and values, individual and collective performance, satisfaction, communication and collaboration within teams);

– professional equality (i.e. equal access to telework, equal treatment, recognition and career development);

– health and quality of life at work (i.e. workload, exposure to occupational and psychosocial risks, quality of professional relationships, work–life balance).

Whether it is led by a project manager or project director within the work organization, or by an external party (e.g. a consultant), different stages can mark the deployment of telework.

7.4.1. *Deploying and experimenting with telework*

When using telework, one of the first steps is to identify the beliefs and expectations about telework (from the point of view of decision-makers, managers and employees). This step involves time for small group discussions combined with plenary meetings to raise awareness and share knowledge and experience. The objective, from the outset of the process, is to deconstruct potential misrepresentations about telework (sources of fear and resistance, as well as idealization) and to co-construct a more objective representation of telework and its impacts. Results from the scientific literature in this field, as well as feedback from previous experiences in similar work structures, can feed into the elaboration of this representation.

Deploying telework also means identifying and involving stakeholders in a sustainable way. In this sense, it is important to integrate the telework deployment approach in the framework of both social dialogue (between employee representatives and management representatives) and professional dialogue (between employees and managers). The creation of a steering committee whose members will ensure the smooth running of the approach is therefore an essential step. The presence of management representatives, human resources players, staff representatives, local managers and employees (representatives of different trades and departments) within the steering committee is essential to promote the decision-making process and the implementation of actions. At the same time, depending on the scope of the project, a community of points of contact or ambassadors may be formed (i.e. resource persons and representatives).

These first two phases, grounded in an organizational, equal and collective approach, are fundamental to establishing a climate of transparency, trust and security from the outset.

Carrying out a diagnosis of teleworkability is an equally important step in the process. The objective is to establish a dialogue and exchanges concerning the positions, the professions as well as the activities that are eligible (or not) for telework. This may involve bringing together representatives of professions, positions and/or fields of activity in order to initiate a reflection integrating the issues of equal access and equal treatment of employees. Indeed, certain positions and/or professions are considered *a priori* as tele-incompatible. There is a risk of exclusion of certain demographics that are better represented in these types of positions or professions. Moreover, as the scientific work in the field underlines, when access to telework is limited, it can be perceived as a social advantage, giving rise to discriminatory behavior and socio-professional exclusion. It can also be penalizing in terms of obtaining promotion or career development in a culture of presenteeism (Gálvez *et al.* 2012; Greer and Payne 2014; Vayre and Pignault 2014). This diagnosis can therefore be an opportunity to identify the opportunities and risks of telework and to define under what conditions it can be a lever for professional equality (and not the reinforcement of professional inequalities).

Following this diagnosis, the conditions of access to telework can be clearly defined and communicated to all members of the organization. Employees will then be in a better position to make a decision about whether or not to engage in this form of work. In this context, members of HR departments and managers will need to have a precise and clear knowledge of the criteria finally selected, the decision-making process and the methods of processing the requests for access to telework which they will receive, as well as of the content and modalities of exchanges with their employees on this subject (a reminder of the conditions of access, eligibility criteria, explanation of the decision taken, medium and long-term perspectives envisaged in the event of initial acceptance or refusal, etc.).

Teleworking arrangements are another important part of the approach. They are the responsibility of each organization and need to be defined collectively, according to individual and collective needs, requirements and constraints, linked to the tasks and activities to be carried out, to the exercise of the profession or to the organization of work. These modalities concern all aspects of telework (duration, frequency and location of the exercise, package types, conditions of reversibility, etc.) and can be fixed or flexible according to the positions, professions or activities concerned.

Once these arrangements have been negotiated, formalizing them is a significant aspect of telework deployment. Indeed, it is recognized that this directly affects the way telework is accepted and perceived by all members of the organization (i.e. social legitimacy (Taskin and Bridoux 2010)). It also influences the establishment of trusting relationships between all employees (whether teleworkers or not) and the development of social control behaviors on the part of managers and teammates (more frequent in cases where telework is non-formal and perceived as having little social legitimacy (Taskin and Bridoux 2010)). This stage involves the mobilization of legal entities and internal communication teams within the organization.

At the same time, the material conditions for the implementation of telework (technological infrastructure, workstation, equipment, tools) must be discussed both from a budgetary point of view and from the point of view of the expected involvement of members of the internal IT services or possible external service providers. Indeed, the appropriate layout of a teleworking space (especially at home), the provision of adequate equipment to carry out all tasks remotely and access to technical assistance help to prevent occupational risks associated with this form of work (Montreuil and Lippel 2003; Mello 2007). The rules for the use of digital technologies and the regulation of these uses can be discussed and (re)defined (e.g. in terms of workload, the right to rest and the right to disconnection (Montreuil and Lippel 2003; Metzger and Cléach 2004; Taskin and Devos 2005; Maruyama *et al.* 2009; Sardeshmukh *et al.* 2012)).

Finally, experimenting with telework is an essential step in its deployment. It takes place, depending on the case, in parallel or downstream of the first stages of the approach. It can be of different scopes and be carried out over varying lengths of time. It may involve experimenting with teleworking with business groups, departments or targeted teams. The choice can also be made to experiment with it more widely within the organization as a whole. The analysis of the activities and uses (real and *in situ*) being deployed makes it possible to identify the benefits and risks of telework, the socio-technical conditions necessary for quality telework, and may help to guide or redirect choices as to the conditions of access, the modalities of telework and the support arrangements for managers and teleworkers. Insofar as the experimentation is collective and joint, and is part of an approach of evaluation and continuous improvement, it is also an opportunity to reassure and secure all the players in the organization (i.e. management and employees).

7.4.2. *Training of teleworkers and managers*

Awareness raising and training of teleworkers is aimed at facilitating a successful transition to telework. Although they are initiated upstream of the change in working methods in order to anticipate the difficulties likely to be encountered when switching to telework very early on, training actions should be extended in the short and medium term, in order to accompany the different stages of the transition.

Mastery of the socio-technical system and the teleworker's appropriation of their new working conditions are among the essential points to be considered. Training content can be proposed that targets the use of the different digital applications and tools made available for working, communicating and collaborating, as well as more broadly targeting the regulation of remote professional relations.

Training that focuses more on self-management (i.e. planning and organizing activities) and the establishment of physical, temporal and psychological boundaries between work and "not work" are equally essential. Indeed, studies show that telework has positive consequences in terms of balance and reconciliation of different areas of life, provided that teleworkers develop skills in terms of the planning and self-management of professional activities and rigorously structure telework periods (Metzger and Cléach 2004; Greer and Payne 2014; Vayre and Pignault 2014). They also stress the importance of the development of a physical work space within the home that is associated with the implementation of behaviors, rules and rituals, in consultation with the personal entourage, which allow the spatial and symbolic delimitation (i.e. role transition) of the work environment within the home (Greer and Payne 2014; Vayre and Pignault 2014).

Managers also play a strategic role in the implementation of telework. They have an important role to play in facilitating the transition to telework, a role which requires new skills and competences. Managers need to manage and evaluate efficiency and remote performance. They must also be attentive and help to maintain or (re)create rich social interactions between teleworkers and their teammates on site, as teleworkers are less exposed to the values and objectives of the organization and often report a lack of face-to-face and informal interactions (e.g. Cooper and Kurland 2002; Hislop and Axtell 2007; Mello 2007; Golden *et al.* 2008; Golden 2009; Sardeshmukh *et al.* 2012; McNaughton *et al.* 2014). The deployment of telework also implies that they are comfortable with the idea of giving their subordinates a certain amount of freedom in organizing their daily work activities and in the progress of the projects and missions entrusted to them. Telework also requires managers to be able to delegate, adopt decentralized work management and develop new performance evaluation and assessment systems (Mello 2007). Training programs for managers are thus aimed at giving them the means to implement new managerial practices, to support them in establishing a sustained and trusting relationship with their teleworkers, and to make them aware of the difficulties encountered by teleworkers, in order to be able to support them in developing strategies for managing their professional activity remotely (Lautsch and Kossek 2011).

7.4.3. *Evaluating the deployment of telework*

In addition to a real-time assessment associated with the telework experiment, a specific system is essential to assess the actions carried out, and also to anticipate those to come. It involves additional information and data collection, through the organization of focus groups, observations of (tele)work situations, individual interviews with target people and/or through questionnaires (quantitative assessment). It aims to report on the experiences lived during the deployment of telework, from the point of view of the teleworkers as well as their managers and colleagues (whether or not they benefitted from telework). The ultimate aim is to determine the impact of telework on individual and collective work and on health and the quality of life at work, in order to consider ways of improving the existing system. The assessment of the approach may also cover financial or environmental aspects.

At the end of the assessment, it is important for the steering committee to define the actions that are part of the extension of the telework deployment approach and which are yet to be carried out. If it turns out that the steering committee is more of a decision-making body then complementary, more operational working groups can be set up. Regular monitoring of the approach, over the long term, will make it possible to take stock of its longer-term impact and to identify the developments that have occurred since the introduction of telework within the organization.

7.5. Conclusion

Preventing the potentially deleterious effects of telework requires a major transformation of the organizational culture. It requires the creation of a real company policy concerning this modality of work (i.e. clear, transparent, voluntarist). It also implies the development of formal arrangements to support telework at institutional, collective and individual levels. Finally, it requires the development of managerial policies and practices (e.g. promotion of autonomy at work, establishment of a relationship of trust), the redefinition of work efficiency and performance evaluation processes and the evolution of the representations conveyed about teleworkers. The creation of spaces for dialogue and collective decision-making is an effective way of achieving these objectives.

7.6. References

Bentley, T.A., Teo, S.T.T., McLeod, L., Tan, F., Bosua, R., and Gloet, M. (2016). The role of organisational support in teleworker wellbeing: A socio-technical systems approach. *Applied Ergonomics*, 52, 207–215.

Biron, M. and van Veldhoven, M. (2016). When control becomes a liability rather than an asset: Comparing home and office days among part-time teleworkers. *Journal of Organizational Behavior*, 37(8), 1317–1337.

Bureau of Labor Statistics (2016). American Time Use Survey: 2015 Results. News release, US Department of Labor, Washington, D.C., USA.

Coenen, M. and Kok, R.A.W. (2014). Workplace flexibility and new product development performance: The role of telework and flexible work schedules. *European Management Journal*, 32(4), 564–576.

Cooper, C. and Kurland, N.B. (2002). Telecommuting, professional isolation and employee development in public and private organizations. *Journal of Organizational Behavior*, 23(4), 511–532.

Duxbury, L. and Halinski, M. (2014). When more is less: An examination of the relationship between hours in telework and role overload. *Work*, 48(1), 91–103.

Gajendra, R.S. and Harrison, D.A. (2007). The good, the bad and the unknown about telecommuting: Meta-analysis of psychological mediators and individual consequences. *Journal of Applied Psychology*, 92(6), 1524–1541.

Gálvez, A., Martínez, M.J., and Pérez, C. (2012). Telework and work-life balance: Some dimensions for organisational change. *Journal of Workplace Rights*, 16(3), 273–297.

Golden, T.D. (2009). Applying technology to work: Toward a better understanding of telework. *Organization Management Journal*, 6(4), 241–250.

Golden, T.D., Veiga, J.F., and Dino, R. (2008). The impact of professional isolation on teleworker job performance and turnover intentions: Does time spent teleworking, interacting face-to-face, or having access to communication-enhancing technology matter? *Journal of Applied Psychology*, 93(6), 1412–1421.

Greer, T.W. and Payne, S.C. (2014). Overcoming telework challenges: Outcomes of successful telework strategies. *The Psychologist-Manager Journal*, 17(2), 87–111.

Hislop, D. and Axtell, C. (2007). The neglect of spatial mobility in contemporary studies of work: The case of telework. *New Technology, Work and Employment*, 22(1), 34–51.

Kelliher, C. and Anderson, D. (2010). Doing more with less? Flexible working practices and the intensification of work. *Human Relations*, 63(1), 83–106.

Lautsch, B.A. and Kossek, E.E. (2011). Managing a blended workforce: Telecommuters and non-telecommuters. *Organizational Dynamics*, 40, 10–17.

Martin, B.H. and MacDonnell, R. (2012). Is telework effective for organizations? A meta-analysis of empirical research on perceptions of telework and organizational outcomes. *Management Research Review*, 35(7), 602–616.

Maruyama, T., Hopkinson, P.G., and James, P.W. (2009). A multivariate analysis of work–life balance outcomes from a large-scale telework programme. *New Technology, Work and Employment*, 24(1), 76–88.

McNaughton, D., Rackensperger, T., Dorn, D., and Wilson, N. (2014). Home is at work and work is at home: Telework and individuals who use augmentative and alternative communication. *Work*, 48(1), 117–126.

Mello, J.A. (2007). Managing telework programs effectively. *Employee Responsibilities and Rights Journal*, 19(4), 247–261.

Metzger, J-L. and Cléach, O. (2004). Le télétravail des cadres : entre suractivité et apprentissage de nouvelles temporalités. *Sociologie du travail*, 46(4), 433–450.

Montreuil, S. and Lippel, K. (2003). Telework and occupational health: A Quebec empirical study and regulatory implications. *Safety Science*, 41(4), 339–358.

Ojala, S. (2011). Supplemental work at home among Finnish wage earners: Involuntary overtime or taking the advantage of flexibility? *Nordic Journal of Working Life Studies*, 1(2), 77–97.

Peters, P., Wetzels, C., and Tijdens, K.G. (2008). Telework: Timesaving or time-consuming? An investigation into actual working hours. *Journal of Interdisciplinary Economics*, 19(4), 421–442.

Sardeshmukh, S.R., Sharma, D., and Golden, T. (2012). Impact of telework on exhaustion and job engagement: A job demands and job resources model. *New Technology, Work and Employment*, 27(3), 193–207.

Solís, M. (2017). Moderators of telework effects on the work-family conflict and on worker performance. *European Journal of Management and Business Economics*, 26(1), 21–34.

Taskin, L. and Bridoux, F.M. (2010). Telework: A challenge to knowledge transfer in organizations. *International Journal of Human Resource Management*, 21(13), 2503–2520.

Taskin, L. and Devos, V. (2005). Paradoxes from the individualization of human resource management: The case of telework. *Journal of Business Ethics*, 62(1), 13–24.

Troup, C. and Rose, J. (2012). Working from home: Do formal or informal telework arrangements provide better work–family outcomes? *Community, Work & Family*, 15(4), 471–486.

Vayre, E. (2019). Les incidences du télétravail sur le travailleur dans les domaines professionnel, familial et social. *Le travail humain*, 82(1), 1–39.

Vayre, E. and Pignault, A. (2014). A systemic approach to interpersonal relationships and activities among French teleworkers. *New Technology, Work and Employment*, 29(2), 177–192.

Vega, R.P., Anderson, A.J., and Kaplan, S.A. (2015). A within-person examination of the effects of telework. *Journal of Business and Psychology*, 30(2), 313–323.

Wilton, R.D., Páez, A., and Scott, D.M. (2011). Why do you care what other people think? A qualitative investigation of social influence and telecommuting. *Transportation Research Part A: Policy and Practice*, 45(4), 269–282

WorldatWork (2011). Telework 2011: A WorldatWork Special Report. Special Report, Scottsdale, Washington, D.C., USA.

8

The Reconfiguration of Managerial Practices through Digital Innovation: The Example of a Work Team in Site Renovation

At the heart of renovation projects and occupational risk prevention activities, this chapter reports on the reconfiguration of managerial practices following the insertion of a business application. After describing what it means to be a renovation manager, we will develop the theoretical axes and the ethnographic approach of the activities which allow us to investigate the object of study. We will show how the business application fits into the activity, the resources it offers, and the constraints it imposes on interactions and managerial relations within a site team.

8.1. Introduction: when digital technology is used on renovation sites

The construction sector is progressively entering into the digital age; the challenge of what is emerging as a cultural revolution, is to make a successful transition from paper to all-digital. The stakes are high, particularly for the management of building sites, since it implies a transformation within a culture that is preferentially based on oral exchanges (often face-to-face) and written media. For the major groups in this sector, successfully making this transition is vital to their long-term survival. It is also strategic for winning new tenders. In addition, meeting the economic imperatives of these new forms of international competition requires a

---

Chapter written by Elodie CHAMBONNIÈRE and Jacqueline VACHERAND-REVEL.

combination of performance, reliability, quality, cost and deadline control in a tight time frame, as well as innovation in work methods and processes. It is at the price of such agility, that lasting capacities for flexibility can be developed that stand out from the competition (Vacherand-Revel 2017a).

In order to understand the motivations of such a cultural disruptive innovation, in individual and collective business practices and the organizational and human impacts of this digital transition, we undertook action research in occupational psychology in collaboration with a company specializing in renovation work. The company was part of an international construction group. Renovation is a specific activity that consists of building on an existing building in order to renovate it (e.g. housing renovations). The objective for the group and the company was to experiment with new business practices – mediated by various digital devices – between the partners involved at all stages of the worksite. The main operators of this digital transformation were a collaborative BIM (Building Information Management) working method based on a shared virtual 3D digital model and business applications.

This research had two aims. The first was to understand the activity of a site supervisor, his interactions with the actors of the renovation sites, and the way his role was likely to be reconfigured under the media coverage of activities. The second aim was to support the digital transition in close association with key players on renovation sites (Chambonnière *et al.* 2018) and to facilitate the use, appropriation and acceptance of digital tools (Bobillier Chaumon and Dubois 2009; Bobillier Chaumon 2013). With these aims in mind, we developed a methodology for the ethnographic investigation of activities (Vacherand-Revel 2017a; Chambonnière *et al.* 2018, 2019). Our research was conducted in real work situations on two renovation sites. One "traditional" site, had traditional digital tools (e.g. for drawing plans) to which we introduced, on an experimental basis, a portable tablet with two business applications – one for assistance in monitoring work and the other for the prevention of occupational hazards. The second "pilot" project used the 3D model and the two business applications.

This chapter will focus on occupational risk prevention, which is not only a major challenge for work teams but also a major area of development for the group. We will focus our attention on the different types of management carried out with teams on worksites, to call, anticipate and deal with risk prevention. We are seeking to understand how the pre-existing management practices of various key site players (manager, site manager and site supervisors) in the occupational risk prevention activity are reconfigured, following the introduction of a business application on one of the investigated sites. We also demonstrate the interest in an ethnographic approach that immerses itself at the heart of this activity, in order to understand the multiple determinants of the evolution of these newly digitized managerial practices.

8.2. At the heart of the renovation sites

To understand the reconfiguration of managerial practices, it is first essential to understand how a renovation project is organized. Their particularities are complex and dynamic systems. At the heart of these systems, the site manager is a true conductor.

8.2.1. *Supervising in a complex and dynamic system*

A construction site is structured into four phases: the response to the call for tenders, preparation, execution of the work and closure. An outside observer might think that they are fenced off; however, the reality of the site is quite different. For example, if a site is poorly prepared, the work will be difficult to carry out and its completion may be delayed. In each of the stages, the tasks that need to be accomplished are interdependent and sequenced. For example, underground networks must be laid before the concrete is poured. For safety first and foremost – and also for the smooth progress of the site – workers are instructed to adopt specific operating procedures so as to not interfere with the activities of others. By sharing the same workplace, two kinds of co-activity exist. Simultaneous co-activity means, for example, that in the same room, a plumber installs the pipe networks while a painter takes care of the walls. Successive co-activity implies that a task carried out at instant *t* by a worker can bring risk to "the one behind".

Life on a construction site is always changing in its configuration. It is therefore marked by the incessant occurrence of more or less serious accidents (water leaks, peeling paint, etc.), that redefine planned actions. Consequently, what marks the working environment on construction sites is its event-driven nature, where each actor must react as quickly as possible to ensure the safety of all. In this sense, different levels of management must work in the company.

8.2.2. *Guiding a worksite: a conductor's activity at the crossroads of various modes of prevention management*

At the heart of construction sites, the site manager is the key player in charge of ensuring that the work is carried out correctly and, above all, safely. As a true conductor, his or her duties include prescribing and coordinating the work of the many construction workers from different building trades, who work together on the site. On a daily basis, the site supervisor's essential mission is to supervise the workers to ensure that they comply with the safety protocols in place (e.g. compliance by wearing personal protective equipment).

Within occupational risk prevention activity, there are several stages of management (we will return to this in the results). First, the site supervisor is responsible for supervising the workers. Then, the site manager, n+1 of the site supervisor, considered as the intermediate management, is in charge of transmitting information concerning safety and is available to the site supervisor for any assistance and feedback. Finally, the operations manager (n+2) and the company director (n+3) have the important responsibility of ensuring compliance with safety protocols.

Teaming up on a construction site is therefore not self-evident if we take into account the plurality of the players involved and the different professions that shape practices and condition social interactions. The analysis of these worksites shows us that their development depends on the confrontation of these different socio-professional worlds. In these, the languages of action and professional visions have been forged, which are not necessarily convergent in the ways of speaking about, describing, understanding events and perceiving a risk situation. These languages, like these "visions", bear witness to a singular experience and the endogenous way of doing things. They reflect the particular interests of the site's partners and the development of specific professional knowledge, built with reference to the organizational and cultural universes in which they evolve on a daily basis (Vacherand-Revel 2017b). In addition, each player has a different "vision" of the world of site prevention. This is not always simple in the management practices carried out on a daily basis – all the more so when inserting a digital innovation. Hence, we have built a specific theoretical framework, adapted to the complexity of renovation sites.

8.3. Understanding occupational risk prevention activity and prevention management

To grasp the amplitude of the changes of this digital transition in the activity, particularly in prevention management, it is necessary to confront them with various levels of complexity and specific dynamics throughout the worksite. Understanding them requires the development of theoretical frameworks capable of taking charge, without isolating *a priori* in the analysis, the organizational and situational elements likely to shed light on the specific function of digital mediatization in management practices. With this intention, only a systemic approach seems relevant to us. In this perspective, we propose a theoretical articulation between the situated and embodied dimensions of the activity and its historical-cultural development (Vacherand-Revel 2017a, 2017b; Chambonnière *et al.* 2018, 2019). This articulation is referred, on the one hand, to *Action Theories* and, on other hand, to *Activity Theories*. It aims to understand the mediatized work as it is carried out in a situation, as it is experienced, tested and developed in its concrete management methods in the occupational risk prevention activity.

Action Theories are rooted in the interactionist and situationist paradigms. For the purposes of this work, we will retain that situationists thematize the indeterminate character of action, emphasizing its nature as emergent, from the dynamics of interactions and contingent, because it depends on the particularities of a given situation. We are attentive to all the modalities used to manage the site, face to face or remotely: verbal, non-verbal, spoken or written modalities. Particular attention is paid to the use of artifacts and the way they are integrated into prevention management, contextualized within the various places and times of action of a worksite.

Activity Theories are rooted in the constructivist paradigm and are inspired by Russian historical-cultural psychology, which is based on the idea of a joint determination of the social and the psyche. According to this orientation, activity does not refer to a circumscribed entity in a defined space-time, nor to a group of people or an organization. The outlines, limits and structure of the activity are defined by its object. What distinguishes *activity systems*, according to Engeström (1987), from one another are their purposes. All activity is directed, that is, oriented by an object. The latter refers both to concerns shared by the collective and to a trajectory oriented towards the product of the activity. The activity is always collective since it involves the coordinated intervention of actors from the building world forming a community by means of an infrastructure of socio-technical mediation (division of labor, rules, norms, instruments), which itself carries a social definition of the productive orientation that constitutes its object.

What interests us in the site situation is what Clot (1999) refers to as the "*reality of the activity*". The activity is "threefold directed: *by* the conduct of the subject, directed *through* the object of the task, it is also directed *towards others*" (Clot 1999, p. 98). Also the *reality* of the activity integrates into visible action, a multitude of potentialities, giving the collective the resources to act collectively or, on the contrary, reasons for *impeding* it by disorganizing it.

The theoretical articulation that we propose sheds light on the density of the activity of preventing professional risks and management practices. It will be observed both in the locality of the multimodal interactions that carry it out, with its contextual backgrounds, in the way the artifacts are grounded in this activity, and in the perspective of its developments and its impediments. We will also examine the way in which this transition from paper to digital (via insertion and the first uses of the business application) is likely to redirect the prevention activity and the types of management in the site team, in what it authorizes, facilitates, constrains or prohibits. The ethnographic approach, described in the following section, is based on the same intention to observe *in situ* the daily work of the site as a whole and its locality.

8.4. Ethnography of the activity on a renovation site

Ethnographically investigating the complex building sector is demanding. It requires time, commitment and a minimum prior understanding of the organization, events and constraints of a building site. These are the conditions to be accepted, to gain the trust of the actors and to build what has become "our" research field. It is also what makes it possible to have an ecological validity of the data collected, an essential point for understanding practices that are not "ideal", but effective in the authentic conditions of their realization. The first period of familiarization, lasting five weeks within the field, enabled us to see and identify what it would be appropriate to observe and to define a place for ourselves. We also conducted informal exploratory interviews with the company's employees. Once this period had elapsed, it was necessary to enter the "representation stage" and obtain privileged access to the "backstage" – in Goffman's sense (1971) – of the worksite, where, on an experimental basis, a business application was being introduced. It should be noted that at the time of our research, the company was in the early stages of its transition to an all-digital worksite.

For a social housing client, the work consisted of transforming an existing building into a 108-unit social residence. During the work, which lasted six months, with two days a week of continuous presence, we immersed ourselves *on* the site. The rest of the time, we were immersed in the company and involved in various strategic projects (digital transition, occupational risk prevention, etc.). The analysis of occupational risk prevention via prevention management could only be relevant if it was part of a significant time frame, from the point of view of the site in question. It was at the price of this prolonged immersion, fostering acculturation and socialization with the practices, that we were able to question, observe and analyze the work that was being carried out. We were given the opportunity to understand the language, the specialized vocabulary of the professions and the implicit terms shared by the actors. We also had to familiarize ourselves with safety instructions and wear personal protective equipment. Without this "equipped" presence on the site and the trust built up with actors over the long term, some of the phenomena observed would have been incomprehensible and others would have remained "invisible" (Star and Strauss 1999).

Carried by these principles and convictions, we *closely* followed the site manager on all of his travels and we were there, *in his shadow*, the privileged witnesses to his interactions and management practices with the actors involved in the project. This specific practice of organizational ethnography is called "shadowing" (McDonald 2005). The first position, in which we were identified as researchers, proved to be difficult to sustain. In fact, because of the particularities of the worksite where we had to act quickly within a defined time frame, we were asked to participate as resource persons in certain tasks and were recognized as credible interlocutors in the situation. With a few skills recognized by the actors involved in the worksite and trust granted, the ethnographic position evolved to take the form of *observant participation* on the worksite *alongside* the site supervisor's "*shadowing*" observation. By taking part in the activity, we socialized and familiarized ourselves and we better understood the activity of the site supervisor and the various types of management revolving around this actor. In addition, as part of a digital change support approach, we were in charge of the proper deployment of the business application: we helped managers, directors and site supervisors to install and use this digital tool. This close contact with the players on the site enabled us to identify the impact of the introduction of this business application on management practices in the occupational risk prevention activity more precisely.

Our research partner, the site supervisor, was 24 years old. At the construction site, he was responsible for coordinating the work of 21 subcontracting partner companies. A maximum of 10 companies (i.e. 35 construction workers) could work together. In addition, 15 professionals from various professions (client, safety prevention, etc.) were involved.

We designed a range of instruments *ad hoc* to the life of a construction site in order to investigate it, while respecting its inherent constraints. Initially, we favored open observations on the site manager, and then targeted observations on activities. Certain moments in the life of the building site were recorded with a camera. The objective was to observe, in detail, the activity *as it was carried out* and the *way it was carried out*. The data collected were ethnographic notes from daily worksite monitoring diaries and transcripts of audio-visual recordings, supported by photographs. It is on this basis that we built our analyses and came up with the results that will be presented. The thematic analyses were co-constructed with the site manager, the operations manager and the safety officer.

8.5. Confirming a culture of safety: prevention management

As a reminder, the prevention of occupational risks was the company's top priority. By analyzing it from a managerial point of view, we demonstrate that prevention was dealt with at three interdependent levels, which stemmed from the same institutional intention while being carried by and directed towards different players. By analyzing the forms it took, we identified management towards site supervision via the managers, then intermediate management via the site manager and, finally, management towards site workers.

8.5.1. *Management towards site supervision*

Once a month for two to three hours, the operations manager conducted site visits, known as “management reviews”, to ensure that site operators understood and respected safety protocols. Accompanied by the site management team (site manager and site supervisor), he visited the site and then reviewed the safety “essentials” that needed to be complied with (e.g. wearing personal protective equipment). During the visits, and before the application was inserted, he used a paper grid where all items were listed (e.g. “the scaffolding procedure”). If they were complied with, he then had to tick “compliant”, “non-compliant” and add a comment if necessary. Once his visit was over, the operations manager sent his report, by e-mail, to the site management team and to the company manager.

In addition, management towards site supervision was carried out indirectly through the presence of an external safety prevention officer. He made “unannounced” site visits to observe the rules and train the site manager in prevention by anticipating risk situations. During the visits, which lasted on average 30 minutes, he made positive “preventive” observations (e.g. “circulation spaces were clear”) or deviations (e.g. “vertical storage of materials was prohibited”). At the end of the visit, the site supervisor needed to correct and respond to deviations (e.g. by storing materials horizontally).

Before the insertion of the business application, when the Safety Preventive Officer made his visits and made an observation, several sequential operations were necessary. First, he took photographs and made an oral comment. Once his visit was completed and he left the site, a digital report was sent to the operations manager by e-mail in PDF format. This document was then forwarded to the site manager, who in turn forwarded it to the site supervisor. In this report, the photographs with the comments were listed. Once the report was received, the site supervisor printed it out to read. He responded to any discrepancies by annotating the report. He then scanned the report and sent it by e-mail to the safety inspector (and copied in the site manager and operations manager), sometimes with photographs attached.

8.5.2. *Middle management*

On a daily basis, the site supervisor worked in constant collaboration with her direct superior: the site manager. For example, she received any new safety information from her management. In the event of serious problems, she could also report them. He regularly toured the worksites to make the site supervisor aware and train her in safety protocols and to ensure that she implemented the appropriate safety devices. Instructions were given orally or by e-mail. The site supervisor was responsible for the progress and proper execution of the work. She needed to act in accordance with the directives of management while assuming sole responsibility for the organization of the life of the construction site. The site supervisor, except for a serious problem, never contacted her management but "*always*" went through her site manager for any requests or feedback. The operations manager expected the site managers to instill a safety culture among their n-1's since they were "*permanently with them on the site*" and "*they were also there to train them and transmit knowledge on the job*".

8.5.3. *Local management: towards construction workers*

The site supervisor needed to ensure the safety of the workers and all of the people going on to the site by preparing, as much as possible, a secure site. The aim was to anticipate risk situations by calling on people to intervene (e.g. safety prevention officer) and by setting up safety devices – for example, defining the operating and intervention methods of workers who worked in a joint activity situation.

In order to guarantee the safety conditions, several paper formats were necessary (list not exhaustive): general safety instructions, emergency services numbers, internal regulations and the Site Installation Plan, called a *Plan d'Installation de Chantier* (PIC). The PIC gathered, on the same document, the information corresponding to the construction site: delivery and storage areas, construction site premises (offices, sanitary facilities, etc.) and safety devices such as the identification of emergency exits and the assembly point. Also, binders contained documents that were essential for construction site life, such as the Special Safety and Protection Plan for Employees (*Plan Particulier de Sécurité et de Protection des Salariés*, PPSPS), which made it possible to identify risks and avoid them during joint activities. There were also the employee identification registers and the welcome documents for construction workers.

In addition, on a daily basis, to anticipate risk situations, the site supervisor conducted site tours to verify that there were no risks. For example, to avoid falls because of "holes", she installed protective barriers. She also ensured that the site was always clean and that the construction workers respected the safety instructions. The cell phone was used by the site supervisor to make calls and take photographs immediately. It enabled her to coordinate and regulate activity on the site to achieve safe conditions. When she was unsure about safety, she could contact the safety prevention officer. To ensure safety conditions on the worksite, photographs were commonly used as proof of a hazardous situation. They were sent by e-mail to the persons concerned, or shown in person on the site, on the site supervisor's cell phone. For example, she took photographs to indicate to the site workers that the circulation spaces needed to be cleared of clutter, so as not to obstruct traffic.

8.6. Digital innovation in occupational risk prevention: restructuring of management practices

The business application was implemented in March 2018 within the company. It was intended for managers, site managers and supervisors and the safety prevention officer. It was used on a tablet and cell phone. A website linked to the business application was used to consult all the reports of site visits and to identify the nature and number of preventive observations noted, as well as the response time issued by the site supervisors. By living at the heart of the company, we observed the beginning of the deployment of this application and identified the first transformations in the occupational risk prevention activity and more particularly, in the related management practices. Let us now examine the different types of preventive visits that this new business application made possible.

8.6.1. *Hierarchical visits by management*

With the insertion of the business application, the visits from management to site management, known as "management reviews", are now called "hierarchical visits". Although the terminology has changed, the sequence of these visits is unchanged. The operations manager completed the list of items, which was constantly evolving, digitally and no longer on paper. At the end of the visit, the business application generated a report that was automatically sent by e-mail to the operations manager.

One notable change was related to the use of the website linked to the application. The reports of hierarchical visits were visible to Group managers and Quality, Safety and Environment (QSE) managers. Research (Bobillier *et al.* 2018) has identified that managers' activities were increasingly transparent with technological tools, which had the corollary of increasing control over their activity.

Managers could use the business application as a panoptic control system (Foucault 1975) by seeing response time. This "digital vision" can also be experienced positively. For example, this was a positive point for the operations manager, since his own superiors saw how close he was to his site teams: "Now, since my name is on every visit, they can see that I make [refers to the visits], so that's good".

8.6.2. *Prevention visits by middle management*

About one year after the business application was inserted, in January 2019, a new prevention action was implemented. Site managers were now asked to make two visits per month to each site. The objective being to reinforce a safety culture as close as possible to construction sites, via middle management, which was in daily contact with the site managers. These visits, which used to be done mainly orally, now needed to be listed in the business application to ensure data traceability. For example, the construction manager who managed three sites now needed to carry out six prevention visits or compliance visits each month. First, it was necessary to write in full, the preventive observations witnessed on site. Second, the preventive observation lists pre-established by the group and company needed to be selected. This was done by checking "yes" (the listed safety device is compliant) or "no" (the listed safety device is not compliant). Generally to save time, site managers chose to carry out compliance visits "since they just had to check boxes instead of writing everything".

Most site managers mentioned that the business application was very easy to use "when they go around the sites with their fellow workers" and that "it is a pretty extraordinary tool", which saves them time, in particular with the pre-established sections enabling quick visits by reviewing the essentials of safety. For example, during the COVID-19 period, lists of items were created by the group and the company to prepare for the resumption of construction sites and to verify that everything was in place for when they restarted. As soon as the sections were established, for example "COVID before resumption" and "COVID visit", the site visits could be carried out, taking into account this new information. On the contrary, there was visibility of the management practices carried out by the site managers, since all their visits had been recorded in the business application.

8.6.3. *Close supervision of the construction workers*

The prevention visits, as was the case with the paper-based system, were carried out by the safety prevention officer. From this point, with the business application, he carried out his tours of worksites with a portable tablet and the notifications acquired were immediately visible for all of those involved. On the business application, he recorded his observations with photographs. Once his visit was complete, the business application generated an e-mail report to the director, operations manager, manager and site supervisor. When the latter received the notification, he connected to the business application to respond to any discrepancies. Reports of prevention visits were also stored in the business application and, as a result, were visible to all. We saw that with scanning, the division of labor was disrupted. The entire company, the QSE manager and the group's senior managers could see the number and nature of observations and the response times issued by the site supervisors to respond to discrepancies.

Our observations show that the use of the business application was very helpful for the site manager. As it provided her with a tool to centralize information, it was easier for her to anticipate risk situations. She could prepare and continuously monitor safety by recording, as a technical aide-memo, the deviations that pointed out the improvements to be made. In addition, her "power to act" was developed in her interactions with her construction workers, since the business application enabled her to issue reminders to better monitor the correction of risky situations. Indeed, most of the time, these deviations implied that the construction workers corrected the points raised. By specifying operating procedures to intervene more precisely, her directives were better followed. For example, masons needed to ensure that guardrails were put in place. With the business application, the site supervisor could have control over construction workers who did not follow safety procedures, as the following verbatim indicates: "it allows me to act, know how long it's been going on, make calls to order". In addition, she was more autonomous since she no longer had to deal with the numerous e-mail transfers. She could now have direct access to the reports. Because of its ease of use in real-life situations, the business application saved time and accuracy in registrations since it enabled a typology to be made, classified by professional risks. Previously, for the site manager, responding to a deviation took about ten minutes, compared to the new time of thirty seconds. The old process involved a large number of time-consuming steps and risks of loss or damage of information. While we know that time manages a large part of the organization of worksites, this gain was far from negligible, especially in the management towards construction workers.

We also identified usage constraints that were being updated in the very constrained environment of the site. For example, it was necessary to think about sufficiently charging the telephone and/or the portable tablet, which may seem trivial, but was not so trivial on a site remote from the site supervisor's office, which did not have sockets. The site supervisor needed to remember to update data and record answers. During our observations, it happened that the site supervisor had to rewrite all her comments because she had forgotten to record them, disturbed by an external risk during her work. We then noted her irritations: "Damn, it didn't record, pff, I have to start all over again". We know that the emergence of emotions due to technology is frequent during the first digital uses (Bobillier Chaumon 2013), which is corroborated by our work.

8.7. Conclusion: towards a better consideration of digital innovations in prevention management

Renovation worksites are marked by three main points: (1) work, most of the time, with subcontractors, which implies the site managers need to ensure that they respect safety; (2) adapting to the existing building by making it safer (e.g. reinforcement of the existing framework); and (3) dealing with incessant risks of human or material order (e.g. adapting the modus operandi of construction workers following an unforeseen event). Digital tools should help improve these three points in the activity of construction managers and save time with regards to activities. It also involves ensuring better communication with subcontracting partner companies, better anticipation of professional risks on site (e.g. modeling using a 3D digital model for better visualization) and finally, knowing how to intervene during incidents to better prevent them.

Focusing on the progress of the worksite, from the point of view of the site supervisor's activity, provides a better understanding of how the introduction of a digital tool helps or hinders the development of an activity such as occupational risk prevention, with the aim of supporting organizational innovation "centered on users and their activities" (Chambonnière *et al.* 2018, 2019). The ethnography of the activities not only allows us to be at the heart of life on worksites, as close as possible to the experience of the players, but also to be included in their games to understand the activities. This approach is essential to support site managers on a daily basis in the digital transition. It was a question of us guiding their learning, helping them in the installation of new digital tools and engaging a dialogue between peers to frame the action, the interactions and the collaborations. The stakes were high, since collaboration could not be decreed; it is always a fragile social construct with rewards, but it also has a human and collective cost (Vacherand-Revel 2007). In this sense, we believe that it is appropriate to discuss the constraints as well as the advantages of their uses and to promote the appropriation of tools, by adapting them to their local needs and managerial practices.

8.8. References

Bobillier Chaumon, M.E. (2013). Conditions d'usage et facteurs d'acceptation des technologies dans l'activité : questions et perspectives pour la psychologie du travail. Memoir HDR Thesis, Université Pierre Mendès-France à Grenoble, France. [Online]. Available at: https://halshs.archives-ouvertes.fr/tel-01559686.

Bobillier Chaumon, M.E. and Dubois, M. (2009). L'adoption des technologies en situation professionnelle : quelles articulations possibles entre acceptabilité et acceptation ? *Le travail humain*, 72(4), 355–382.

Bobillier Chaumon, M.E., Cuvillier, B., Sarnin, P., and Vacherand-Revel, J. (2018). Usage des TIC et évolutions des pratiques socioprofessionnelles des cadres : quels repères pour le métier et quelles incidences sur la santé ? *Pratiques psychologiques*, 24(4), 349–373.

Chambonnière, E., Vacherand-Revel, J., and Andrieu, B. (2018). Transição digital na construção civil: analisar a atividade de "levantamento de inconformidades" mediada por um aplicativo informático de uma diretora de obra. *Laboreal*, 14(2), 45–61.

Chambonnière, E., Vacherand-Revel, J., and Andrieu, B. (2019). Innovation numérique pour la prévention des risques sur un chantier de réhabilitation. *Actes du 54ème Congrès International de la Société d'Ergonomie de Langue Française (SELF)*, 149–155.

Clot, Y. (1999). *La fonction psychologique du travail*. Presses Universitaires de France, Paris, France.

Engeström, Y. (1987). *Learning by Expanding: An Activity-theoretical Approach to Developmental Research*. Orienta-Konsult, Helsinki, Finland.

Foucault, M. (1975). *Surveiller et punir*. Gallimard, Paris, France.

Goffman, E. (1971). *Relations in Public*. Basic Books, New York, USA.

McDonald, S. (2005). Studying actions in context: A qualitative shadowing method for organizational research. *Qualitative Research*, 5(4), 455–473.

Star, S.L. and Strauss, A. (1999). Layers of silence, arenas of voice: The ecology of invisible work. *Computer Supported Cooperative Work (CSCW)*, 8(1–2), 9–30.

Vacherand-Revel, J. (2007). Enjeux de la médiatisation du travail coopératif distribué dans les équipes de projets de conception. *Perspectives interdisciplinaires sur le travail et la santé*, 9(2), 1–2.

Vacherand-Revel, J. (2017a). Le travail coopératif d'équipes de projet d'ingénierie à l'épreuve de l'activité en réunion médiatisée et multi-localisée. *Psychologie du travail et des organisations*, 23(2), 89-116.

Vacherand-Revel, J. (2017b). Le travail en équipe à l'ère de sa médiatisation numérique : figures actuelles de l'innovation organisationnelle. *Psychologie du travail et des organisations*, 2(23), 85–88.

9

Integrating Collaborative Robotics into Work Situations: The Intentions of SME Managers in the Digital Transformation of their Companies

This chapter focuses on the development of the digital transformation intention of SMEs by their managers, through the case of collaborative robotics. Based on a dialogical model of the design processes, this work takes into account the importance of the consideration of "reality" by the managers, to shape their transformation "desires": in the perspectives they mobilize, the managers will consider material and social circumstances of their company and not only have the goal of controlling productivity.

9.1. Transformations in work situations seen through the prism of technocentric solutions

The work presented in this chapter deals with one of the questions of the interdisciplinary research project ANR[1] HECTTOR[2] which focuses on work transformations related to collaborative robotics technologies in small and

---

Chapter written by Anne-Cécile LAFEUILLADE, Flore BARCELLINI, Willy BUCHMANN and Tahar-Hakim BENCHEKROUN.

1 *Agence Nationale de la Recherche* (French National Research Agency).

2 *L'Humain Engagé par la Cobotique et les Transformations du Travail et des Organisations dans les usines du futur*, http://crtd.cnam.fr/hecttor-un-projet-anr-sur-la-robotique-collaborative-et-l-industrie-du-futur-dans-les-pme-998785.kjsp.

medium-sized enterprises (SMEs). To date, collaborative robotics appear to be very emblematic of the digital transitions towards an "Industry of the Future".

In the French context, the "Industry of the Future" program aims to encourage industrial companies – particularly SMEs which, for the most part, are the fabric of French industry – to modernize their production apparatus through technological and organizational changes. It is in response to an observation made by the *États Généraux de l'Industrie* (French General States of Industry), which pointed out that France has a deficit, in terms of investment, compared to its main competitors. French companies are slow to renew their equipment and invest mainly because of wear and tear or aging of their production equipment (Bidet-Mayer 2016). The program therefore urges managers to modernize their companies through a transformation based on the introduction of new technologies, associated with a lean management-type work organization. However, analysis of the program's content shows that little consideration is given to the actual work, and the historical and social characteristics specific to SMEs; as well as to the change management processes implemented by managers in these "digital transformations" (Lafeuillade *et al.* 2019). It is on this last point that this chapter focuses.

To deal with the question of how "digital" transformations of SMEs are carried out, we follow research on the activity of SME managers (Mintzberg 1984; Duchéneaut 1998; Ghram 2009), who are considered to be guarantors of: (1) transformations in the productive tools of SMEs and (2) opportunities for "healthy" work and worker performance. We consider a manager to be any actor involved in the management of the company, in addition to the site manager, with hierarchical power and responsibilities over one of the company's entities, for example an industrial manager or a production manager. Depending on the size and organization of the SME, it is possible for a manager to combine these management functions.

We also consider the management of transformations related to the introduction of collaborative robots. In a context where the message of modernization seems to be highly technocentric, collaborative robotics is presented as a solution to address some of the physical constraints of operators' work. Ranging from collaborative robotics[3] to *exoskeletal solutions*[4], collaborative robotics technologies are distinguished from conventional robotics by the absence of physical barriers, separating the industrial robot from the operator. They are designed to intervene when the operator is confronted with tasks that are arduous, dangerous, repetitive or with "very little added value" but which remain indispensable (Kleinpeter 2015). These technologies tend to appeal to managers because they are presented as

3 http://www.inrs.fr/risques/robots-collaboratifs/de-quoi-parle-t-on.html.

4 http://www.inrs.fr/actualites/brochures-exosquelettes-travail.html.

"solutions to musculoskeletal disorders" that are inexpensive and "easy" to integrate and maintain.

Finally – following the work in activity-centered ergonomics–we consider that transforming the existing technology for a specific purpose (here, integrating collaborative robotics as a new technological means of work) can be seen as a process of designing a future work situation (e.g. Daniellou 1994). In this framework, we seek to characterize the key elements on which managers rely to develop their intention to equip their production tool with collaborative robotics technologies.

In this chapter, we will first present a brief theoretical framework on management activity and design processes, followed by our methodology for data collection and analysis, and finally, the results, which shed light on the equipment intentions of managers of five French SMEs.

9.2. Models of leadership activity to understand change management processes

A review of the literature makes it possible to present several models of what the activity of SME managers could be for engaging their company in a transformation project.

9.2.1. *The activity of managers at the crossroads of different roles*

The first, rather prescriptive, models view the SME manager as an "entrepreneur" (Duchéneaut 2008) who "takes risks" (Knight 1971; Marchesnay and Julien 1987; Laurent 1989) to transform his/her business and who will commit existing resources to economic activity for a probable future outcome (Drucker 1985). In this model, we also find the approach of Schumpeter (1934) for whom the entrepreneur is the one who invests, innovates, changes production techniques and discovers new markets. In the case of SMEs, this role is reinforced: the manager must be responsible for many tasks and must possess multiple skills, unlike large organizational structures that are staffed by specialists. Moreover, their relationship with "power" – or exercising their function – depends on their "profile". This profile is constructed according to several factors: the origin of their position as a manager (creation, takeover or family), training (from self-taught to graduate), management "logic", motivations and ambitions, their way of working and the interweaving of the private and professional spheres (Duchéneaut 2008).

These approaches characterize the predominant role of managers within a company; however, they say little about what their practices might be, particularly in change management processes.

If one looks more closely at what a manager actually does, Mintzberg's (1984) work accounts for the "roles" performed by managers. Some roles refer to relationships with others: the contact, the leader and the icon. Other roles relate to information: the disseminator, the observer and the spokesperson. In addition, managers take on roles related to decision-making and change management: the entrepreneur (in the Schumpeterian sense), the resource allocator, the regulator and the negotiator. Finally, the manager is also the one who develops arguments from working situations to strengthen their power to act (Ghram *et al.* 2009).

9.2.2. *Developing the intentions of managers in change management processes: the contribution of the dialogical model of design*

Previous models do not always reflect the manager's activity, as seen in activity-centered ergonomics. In this approach, the activity is a finalized construction carried out within a framework, that is, it is carried out with regard to the material, social and historical components of its context (Daniellou and Rabardel 2005). This framework includes tasks, means, organization, etc. and is built around several perspectives (economic, social, historical, cultural, etc.); it is itself constantly revisited and reinvested in by the subject's life experience. In this sense, the activity of managers is necessarily situated in a given context (Béguin 2007).

To try to understand the activity of managers who are undertaking a transformation of their company's production equipment, we propose to see this activity as part of the conduct of a design project initiated by a technological equipment intention. To understand the managers' activity is thus to characterize, initially, the elaboration of this intention in the given context in which the manager is located.

To do so, we rely on the models of design project management proposed by activity-centered ergonomics, which sees design as "a process of progressive determination of an object or work situation and the actions of the actors in the process" (Terssac and Friedberg 1996, cited by Béguin 2004, p. 5). In this sense, design is characterized by a process that takes place at two distinct levels: on the one hand, the goal to be pursued, which is the "project", and on the other hand, the transformations needed to achieve the goal, which is the "project management". The "project" corresponds to the purpose pursued, to the desire that relates to the future (Daniellou 2004); in other words, to the intentions carried by the decision-making actors of an organization – in our case, the managers of SMEs. "Project management" is the concretization of these intentions, which requires the production of several "draft" solutions to the project. Managers are then actors in the construction of the intentions and the structuring of the project to transform said intentions into concrete achievements. This involves, in particular, making trade-offs beween financial, temporal or human resource issues related to the project. These actors, who bear the intentions, will also mobilize the various – often technical – teams to be involved in the project, who will be in charge of assessing the *feasibility* of the project in relation to the intention (Daniellou 1994). The use of professional design skills (industrial engineering, computer science, etc.), organization and training, whether provided by the managers or not, makes it possible to specify the project and propose potential design solutions.

The structure of the design process should allow for this discussion between the development of the intention and the search for feasibility (Martin 2000); objectives can be rethought, for example in the case of technical characteristics that are difficult to achieve. The "project" and "project management" can thus be seen as a discussion between the "*desirable*" and the "*possible*" (Béguin 2004) or between the "virtual" – the intention – and the "reality" – the feasible and the possible. The iteration that takes place between these dimensions of design originates in the "*conversation with the situation*" (Schön 1987): the actors aim at a design, and they project ideas and knowledge, but the "situation" responds to them by presenting unexpected resistance that forces them to pose the problem differently.

In addition to this model, previous research has highlighted the pitfalls of project management, in particular: an intention structured mainly by technical or financial perspectives, a potential disengagement of the managers in the conduct of the project which is led by technology or finance – the technical or financial feasibility then dominating design decisions to the detriment of a more multi-perspective vision (technical, financial, health, safety, performance, human resources, organizational) for the project in relation to the initial intention – and finally, a flawed and theoretical model of the work and these transformations carried by the actors of the design, whether they be the managers carrying the intention or the designers carrying the feasibility (Barcellini *et al.* 2014). The *implicit* functioning of the human being at work is often the "human factors" model, which defines limits in working conditions; however, it does not sufficiently consider the work situation as a whole.

Together, these works allow us to characterize the development of the transformative intent of SME managers. In the following, we propose some elements of understanding and framing:

– The managers are among the actors of the design; they are the bearers of the *desire relative to the future* (of the *desirable*) and/or of the *feasibility*. In the case of SMEs, the same actors can potentially carry these two dimensions, the distinction is made – or not – according to their hierarchical status and their knowledge of the technologies and management methods of the "Industry of the Future".

– We consider here the *desire* as being constituted by the elements of the context that the manager retains as relevant in the construction of the project. What is retained as relevant is singular, specific to each manager; managers therefore each have *their own desire.*

– The context of the managers' activity considered is that of the company in its technical, social and organizational dimensions. This context plays an important role in the development of the intention to introduce new technology. It is then a question of understanding which elements of this context – of reality – are mobilized by the managers in order to build their intention.

– The development of an intention is potentially based on the mobilization of several perspectives (financial, technical, health, safety, working conditions, HR, etc.). The aim here is to understand which perspectives are mobilized by managers to build their intention, particularly those related to their representation of work, and how these perspectives are built, fed or impacted.

– Finally, in order to investigate the question of the *desire* of managers more precisely, we propose two inductive categories to characterize their context of activity:

1) This comes from the intrinsic functioning of the company: the company's infrastructure, the "difficulty" at a workstation or political choices.

2) This comes from the "Industry of the Future" context and collaborative robotics technologies.

Taking these elements into account, we propose the following two research questions to characterize the development of the transformation intention held by SME managers: (1) What elements are chosen retained by the managers to build *their desire*? (2) What are the elements that the managers evoke when this *desire* is "put into discussion" with reality?

9.3. Methodology for data collection and analysis

The field research was carried out at five different companies (see Figure 9.1).

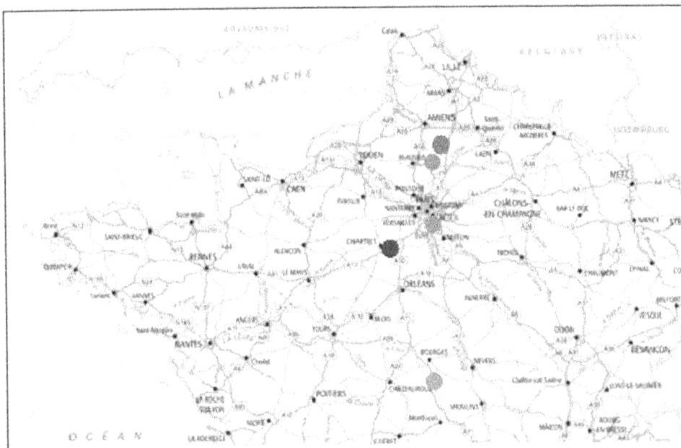

Electronic cabling of cards
120 staff
More than 40 years old

Manufacture of hardware products
70 staff
100 years old

Manufacture of heating tools
55 staff
More than 40 years old

Manufacture of electro-extrusion materials
40 staff
More than 100 years old

Manufacture of ceramic cores
100 staff
More than 140 years old

SME 1: a company manufacturing electronic card wiring, employing 120 people, located in the Hauts-de-France region and existing for more than 40 years

SME 2: a company manufacturing industrial hardware, employing 70 people, located in the Hauts-de-France region and existing for 100 years

SME 3: a company manufacturing heating tools, employing 55 people, located in the Île-de-France region and existing for more than 40 years

SME 4: a company manufacturing steel materials using electro-extrusion technology, employing around 40 people, located in the Centre Region and existing for more than 100 years

SME 5: a company manufacturing ceramic cores, employing about 100 people, located in the Centre Region and existing for more than 140 years

Figure 9.1. *Field of activity, payroll, duration of existence and location of the five SMEs. For a color version of this figure, see www.iste.co.uk/bobillierchaumon/digital.zip*

In four establishments, the research was conducted in a similar manner, namely: over the course of half a day, a semi-structured interview with the managers, followed by a tour of the production site, possibly including a stop at the targeted workstation. These first four visits were carried out by two or three researchers (ergonomics and sociology) between March 2018 and January 2019.

The visit to the fifth company (manufacturing industrial hardware) was conducted using a different framework: the ergonomics researcher was able to join, as an observer, on a day dedicated to the restitution of an on-site needs assessment, commissioned by the company, from the expert in collaborative robotics integration at Cétim[5], Hauts de France. This analysis uses a diagnostic tool specific to Cétim, whose purpose is to help the company to formalize its technological needs to improve its industrial process.

While these five companies seemed to be part of a "classic" Taylorian organizational model, they were nevertheless differentiated by the variability of their transformation intention; in this sense, three configurations were identified at the time of our visits:

1) an intention to provide equipment without a targeted workstation (SMEs 3 and 4);

2) an intention to provide equipment with a targeted workstation (SMEs 1 and 5);

3) an intention to provide equipment enhanced by the experience of initial trials of integrating the technology into the production tool (SME 2).

The objectives of the semi-structured interviews were to gather evidence of managers' interest in collaborative robotics technologies. Three questions – constructed with our sociologist colleagues – guided these interviews. Question 1 (*Why would collaborative robotics be of interest with regard to the questions and issues that concern you about the functioning of your company?*) aimed at gathering the advantage (and the limitations) for the manager of collaborative robotics with regard to the production context of the SME. Question 2 (*How did this intention to implement a collaborative robotics application in your company come about?*) aimed at specifying the genesis of the project to set up such a solution. Question 3 (*Have you already identified a position/workstation? And why? Have you carried out tests?*) focused on the workstations that could be subject to a cobotization, the criteria for choosing potential workstations, as well as their history and context.

Each interview lasted about 90 minutes. All of them were subject to written reports and then an inductive categorization, separating the elements of the intrinsic functioning of the SME, from those of the "Industry of the Future" context, including collaborative robotics.

5 *Centre technique des industries mécaniques* (French technical center for mechanical industries) whose vocation is to promote the link between scientific research and industry.

9.4. Managers' desires in the face of reality: an encounter that helped to shape their intentions

9.4.1. *Elements shaping managers' desires*

The following are the elements chosen by the managers to shape their desires emanating from: (1) the intrinsic functioning of the company, and (2) the context of the "Industry of the Future". In both categories, the elements reflect the mobilization of different logics underlying the development of the desire.

In the intrinsic functioning of the company, reference was made to:

– "arduous" characteristics of a workstation: "we would like to equip finishing operators with a kind of sensor so that they stop working when they reach their own level of fatigue" (SME 5); "the afternoon operator would have to install it (authors' note: the collaborative robotics mobile unit) at the end of the shift and she works at night", "we would like a robotic arm to put on the hoods and offer operators something else more interesting to do" (SME 1); these characteristics are linked to an occupational health logic;

– characteristics stemming from the company's infrastructure or the development of the products manufactured: "we have an old infrastructure that requires a lot of handling and our manufactured parts are increasingly heavy" (SME 4); characteristics linked to a technical logic;

– characteristics stemming from an economic strategy of relocating jobs from low-wage countries: "we have problems with the quality of our products at our Shanghai site, which forces us to review the current organization of production and of our workforce" (SME 3); this refers to a strategic logic.

The desires of the managers, in the context of the "Industry of the Future", revealed their expectations of clarifying the technological possibilities of collaborative robotics. This desire seemed to be based on a prospective logic of research and development, which relied on:

– technological foresight (participation in trade fairs or training courses offered by organizations such as regional Chambers of Commerce and Industry (CCI): "we find out what exists and we work with the CCI, but between science fiction and reality, we don't really know what is possible" (SME 5);

– groupings of professionals led by experts: project-carrying unions associated with partner organizations such as Cétim and engineering schools (e.g. with

Ensam[6]): "We work a lot with Cétim and our professional union on technical innovation projects" (SME 1).

9.4.2. *The "conversation" between the desire and reality*

In a second stage, when what is desirable is "put into discussion" with reality, the managers mentioned: (1) the economic cost of occupational health related to occupational strain; (2) the economic cost of safety to be implemented with the future technical system; and (3) the influence of circumstances arising from past choices or events. These initial "conversations between desire and reality" suggest the importance of the material and social circumstances of the "situation", in our case the SME, which interact in the development of the managers' intentions.

The economic cost of occupational health is reported through the occupational prevention account; among the various risk factors selected by Carsat,[7] night work and repetitive work are frequently encountered in the industrial world. They have an impact on the cost of the company's social security charges. Collaborative robotics, by promising to take over part of the arduousness at a workstation, can become an administrative and regulatory asset: in other words, if a so-called "arduous" workstation is equipped with this technology, it will be less impacted by the financial compensation required from the employer: "the afternoon operator would have to install it (editor's note: the mobile collaborative robotics unit) at the end of the shift and she works at night ... and that way we'd once again have production at night at lower cost" (SME 2); "tomorrow there will be an additional tax on arduousness and strain on the operators, so if these technologies can relieve the operators and show that we are good performers, then it would be a win-win situation for everyone" (SME 5).

The economic cost of the safety to be implemented with the future technical device is referred to through standards; safety standards have evolved with the possible sharing of space between humans and robots, conferred by collaborative robotics technologies. However, each new device must be certified from a safety point of view. Managers will have choices to make in order to achieve these certifications, either to train staff internally or to use an external service provider. In both cases, concrete issues emerge:

– the responsibility to be assumed if the certification is carried out internally;

6 *École Arts et Métiers* (formerly *École nationale supérieure d'arts et métiers, ENSAM*).

7 *Caisse d'Assurance Retraite et de Santé au Travail* (French Retirement and Occupational Health Insurance Fund).

– the negative economic impact on the return on investment if it is outsourced, as the costs of the providers are high.

"We tried on two workstations (authors' note: to integrate a collaborative robot), but security is a problem because it is expensive" (SME 2).

Finally, the development of the intention was linked to certain past choices or events that are part of the history of the SME, for example:

a) The history of the technological equipment of a workstation: "before, the operators used to put the covers on the parts to be protected, then they did the varnishing by hand, then we put a varnishing machine, and now we would like a robotic arm to put the sensors and propose something else more interesting" (SME 1).

b) The experience of a past fact that influences a current choice: "we tried two positions, but safety is a problem, because it is expensive and there is no question of doing it internally, we have already had an accident and the lawyers know their job well" (SME 2).

c) The cost of salaries. This difficulty, clearly explained by the manager of the heating tool company, led him/her to approach technology as a means of bringing back the relocated workforce to the territory at a "lower" cost, which reflects the "robocalize" discourse[8]. Indeed, the major human resource management strategies of previous decades have resulted in a certain relocation of low-skilled labor to economically more advantageous countries; the counterpart is quality management, which can be undermined by the physical distance between the design office, R&D and the production site (Fontagné *et al.* 2005).

9.5. The reality, a messenger from the past, in a modernization project

This research is part of an understanding of change management processes linked to technological equipment, from the point of view of SME managers. The results shed light on the elements that influence the design process, from the point of view of managers. These results come from the discussion between the *desire* expressed by managers to transform their productive tool and the *possible* of the real situation; they show that the intention is developed with the mobilization of several logics, particularly financial, health and safety.

8 Speech at the initiative of Symop and Cétim, 2006: "robot-iser pour ne pas délo-caliser".

These results seem to be in line with the rather classic paradigm of productivity control[9]; the perspective involved refers to a form of economic research into the means implemented for technological solutions.

Nevertheless, circumstances arising from past choices or events, and which are part of the history of the SME, allow other dimensions (or perspectives) to emerge: promoting "interest" in a job, career management in the context of security or human resources management, by looking for the "rise in skills".

The perspectives mobilized by managers are also constructed from choices, rules, events, etc. that have been made (or have taken place) in the past. In this sense, the context of managers' activity bears – in its technical, social and organizational dimensions – traces of previous situations (Lafeuillade *et al.* 2018). This perspective leads us to question the impact of the past (Gaudart 2013), as a determining potential of the design activity of SME managers seeking to modernize their production tool through technological and organizational changes.

9.6. References

Barcellini, F., Van Belleghem, L., and Daniellou, F. (2014). Design projects as opportunities for the development of activites. In *Constructive Ergonomics*, Falzon, P. (ed.), Taylor and Francis, New York, USA.

Bidet-Mayer, T. (2016). *L'industrie du futur: une compétition mondiale*. Presses des Mines, Paris.

Béguin, P. (2004). Monde, version des mondes et monde commun. Texte préliminaire. *Bulletin de Psychologie*.

Béguin, P. (2007). Prendre en compte l'activité de travail pour concevoir. *@ctivités*, 4(2), 107–14 [Online]. Available at: http:// www.activtes.org/v4n2/v4n2.pdf.

Daniellou, F. (1994). L'ergonome et les acteurs de la conception. *Communication au XXX Congrès de la SELF*, 27–32 Paris.

Daniellou, F. (2004). L'ergonomie dans la conduite de projets de conception de systèmes de travail. *Ergonomie*, 359–373.

Daniellou, F. and Rabardel, P. (2005). Activity-oriented approaches to ergonomics: Some traditions and communities. *Theoretical Issues in Ergonomics Science*, 6(5), 353–357.

Drucker, P. (1985). *Les Entrepreneurs*. L'Expansion Hachette Lattès, Paris.

9 One technology replaces another when it performs better (in the more productive sense) (Orivel, 1991). Controlling productivity means accurately determining production times and organizational rules.

Ducheneaut, B. (1998). *Les dirigeants de PME. Enquête chiffres analyse. pour mieux les connaître*. Éditions Maxima Laurent du Mesnil, Paris

Fontagné, L., Lorenzi, J.H., Artus, P., Fayolle, J., and Aubert, P. (2005). *Désindustrialisation, Délocalisations*. La Documentation française, Paris.

Gaudart, C. (2013). *Age et travail à la croisée des temporalités – l'activité face aux temps*. Document pour l'habilitation à diriger des recherches, Université Victor-Segalen Bordeaux, Bordeaux.

Ghram, R., Fournier, C., Benchekroun, T.H., Khalfallah, T., and Six, F. (2009). Strategic decision making: Work analysis of a company manager. *17th Congress of IEA*, Beijing.

Julien, P.A. and Marchesnay, M. (1987). *La petite entreprise – principes d'économie et de gestion*. Vuibert/G. Vermette, Paris.

Kleinpeter, E. (2015). Le cobot, la coopération entre l'utilisateur et la machine. *Multitudes* 2015/1(58), 70–75.

Knight, F. (1971). *Risk, Uncertainty and Profit*. University of Chicago Press, Chicago.

Lafeuillade, A.-C., Buchmann, W., and Barcellini, F. (2018). Quelles opportunités de performance et de développement l'intégration de la robotique collaborative apporte-t-elle aux opérateurs ? Cas des PME françaises. *Actes du 53ème Congrès de la SELF*, Bordeaux, 3–5 October.

Lafeuillade, A.-C., Buchmann, W., and Barcellini, F. (2019). Industrie du futur et PME. Le travail, la robotique collaborative et les dirigeants d'entreprises : premiers constats. *Acte Epique 2019*. Editions Arpège, Paris.

Laurent, P. (1989). L'entrepreneur dans la pensée économique. *Revue internationale PME*, 2(1).

Martin, C. (2000). *Maîtrise d'ouvrage, maîtrise d'oeuvre, construire un vrai dialogue*. Octarès Editions, Toulouse.

Mintzberg, H. (1984). *Le management au quotidien, les 10 rôles du cadre*. Les Editions d'Organisation, Paris.

Orivel, F. (1991). Analyse économique des nouvelles technologies éducatives. *La Formation Multimédia (Actes du Colloque)*.

Schön, D. (1987). *Educating the Reflective Practitioner*. Jossey Bass, San Francisco, USA.

Schumpeter, J. (1934). *The Theory of Economic Development*. Havard University Press, Cambridge, USA.

10

The Role and Function of Technological Artifacts in Entrepreneurial Activity

Several works present the role of artifacts in the cooperation and coordination activities of localized or multi-localized work teams. In entrepreneurial activity, moments of cooperation act as spaces of intersubjectivity in which different objects come into play. In this chapter, we question the status and paths of technological artifacts in entrepreneurial and team innovation activities. Based on a comprehensive and situated methodological protocol, our analyses reveal three processes: the process of creating the technological artifact, which becomes an intermediate object in the activity; the process of developing this intermediate object, which takes on several forms in the activity and performs different functions within the entrepreneurial team project; and, finally, the process of moving from intermediate object status to boundary object status.

10.1. Introduction

Creativity represents a major challenge for entrepreneurial teams that seek to innovate. Among the works on innovative entrepreneurship, some have focused, more specifically, on the role of entrepreneurial devices (Bouchard and Bos 2006) and artifacts in business creation and innovation activities (Poidi *et al.* 2016). This work highlights the importance of artifacts and environments in the development of project team trajectories. Other work presents the role of technological and telework artifacts in the cooperation and coordination activities of localized (Ianeva *et al.*

Chapter written by Irène POIDI, Marc-Eric BOBILLIER CHAUMON and Jacqueline VACHERAND-REVEL.

2017) or multi-localized (Vacherand-Revel 2017) work teams. In entrepreneurial activity, moments of cooperation act as spaces of intersubjectivity (Zarifian 1996) characterized by the meeting of different points of view, different cultures and different social worlds (Star and Griesemer 1989). In these moments of cooperation, technological artifacts and work spaces, both physical and social, participate in the interactions and construction of the work collective (Clot 1999, 2008). These artifacts in entrepreneurial activity can undergo situational transformations and become instruments for the activity (Rabardel 1995; Folcher and Rabardel 2004). They can also take on the status of intermediate objects in the networks of relationships between actors (Vinck 1999, 2009). Finally, these objects can also constitute boundary objects (Star and Griesemer 1989; Star 2010) when actors integrate conventional metadata (Vinck 2009), allowing them to be mobilized by different social worlds. This is particularly the case with intranets that are used by different departments within the same company, each interpreting the role of this tool from a different point of view. These intermediate technological objects and boundary objects will be the focus of this chapter. On the one hand, we seek to define their construction trajectory, and also to determine the precise role played by these objects in the genesis and development of entrepreneurial trajectories. How do these artifacts participate in the collective activity of innovation and creativity within the team project?

10.2. Theoretical foundations

We consider the status of an artifact as an evolving and developing element in the business. Through use, the functions of artifacts are made and undone, constructed and deconstructed. The intermediate function of artifacts and their role in work teams for communication, design and creative activities will be discussed first. We will then address the role of boundary objects in the exchanges and articulations between different actors or different teams, each with their own cultures and modes of operation.

In work collectives, exposure is crucial for the coordination and organization of teams, especially when the activity is widely distributed. Vinck (1999, 2009) refers to any artifact that is mobilized in the networks of relations between actors as an intermediary object. The intermediary object is constructed by the entrepreneurial team for the exposure of exchanges and communication within the team, as well as for design and creativity. This work is focused on the role of objects in team design processes, particularly the design of innovative objects with high added value.

Innovation issues translate into a need for the team to create together, mobilizing various skills, as well as to create quickly, using efficient working methods. This is where joint work supports, which can be material or digital, and which participate in the co-construction of the innovation object, come into play. The intermediate object then fulfills several functions:

1) It is a tool for social expression when it is mobilized in discussions between team members.

2) It is also a project consolidation tool. Lavoisy and Vinck (2000), working on the role of a design support tool (software) within a design office, showed that the role of this tool was not simply limited to assistance and design, but that "through this tool, the identity of the company, the training policy and the philosophy of the design work are articulated". In the work presented by Lavoisy and Vinck (2000), this process of constructing the team's identity took place in three stages: (1) The first stage was that of suspicion, where the tool was not fully integrated into the team's mode of operation. (2) The second stage was the institutionalization of the project. In this stage, an important element is the confidentiality of the project that was materialized by the software tool and the documents printed from it. (3) Finally, the third stage was the centering of the project into an object. Lavoisy and Vinck (2000) show that constructing the object is the same as building the company for the team members. There is a process of substituting the tool for the company. It is a symbolic process of attributing meaning. The main objectives and values constructed for the company will over time be embodied in the design tool.

3) Another function fulfilled by the intermediate object is its supportive role for creativity. Work in ergonomics (Buisine 2013; Bonnardel 2014; Safin 2015) shows the importance of creativite methods in the efficiency of co-design and team innovation work.

4) The intermediate object becomes a process stabilization object. In a single moment, the actors center around the development of the object on which the design will be produced and reproduced. This object centering phase is important in team creativity processes. It allows the teams to validate the selected ideas and to evaluate them through a first design phase. The intermediate object itself constitutes a collective validation of the project by the team members, at a moment of the project. It is both an indicator and a provisional result, enabling the project to move towards the next stage.

5) Finally, the intermediate object becomes a tool for design and communication within the team, as well as a translation tool, when it is directed to people outside the team. This third level is in line with the concept of the social expression tool developed by Vinck (2009), according to which actors mobilize intermediate media to interact with each other and also with actors outside the activity.

In the following paragraph, we will look at the role of objects in interactions between actors who come from different backgrounds and various fields of competence, but who act together in the same process of creativity and innovation.

The boundary object corresponds to what flows between several communities of practice and meets the informational requirements of each of them. It is an object that crosses social worlds or boundaries while remaining constant (Bowker and Star 1999). It enables communication and articulation between different social modes in order to serve a common object (Star and Grisemer 1989; Star 2010). Trompette and Vinck (2009) described the boundary function as the process by which actors from different social, cultural and identity worlds are called upon to cooperate, manage and coordinate with each other, despite their different points of view. The boundary object highlights the role of all kinds of artifacts: directories, classifications, materialized representations (maps, drawings), standardized methods: involved in the collection, management and coordination of distributed knowledge. The role of boundary object can even be played by an individual, a point of contact within a team, in charge of communication and negotiation. The development of the object's intermediate function towards a boundary object function is possible as soon as the actors integrate conventional metadata (Vinck 2009), allowing them to be understood by social worlds different from the design team. Metadata can be rules, professional jargon, domain-specific schematization (graphs, tables), a dress code and other social codes. Metadata can be elements common to two communities or elements acquired by a community to facilitate exchanges with another community.

Through the status of intermediate object and boundary object, the object acquires a stabilized role within the team. It is used in interactions within the team, and in interactions with other teams or other actors outside the team. The boundary object is, in fact, an object that crosses social worlds and allows the inclusion of a greater number of actors in the design.

10.3. Methodology

Our research protocol was based on an ethnographic approach to collecting developmental and local data. Our analyses focused on the activity of an entrepreneurial team made up of a total of seven actors, which was observed over a period of 18 months (in the years 2015–2016). We will call this team "Etui-Innovant" for confidentiality reasons. Table 10.1 summarizes the data collection and analysis protocol.

Data collection methods and specificities of the collection	Methods of analysis	Activities, data and processes identified
Research notebook/research notes: longitudinal tracking of activity (over 14 months)	Developmental trajectory analysis	Formalization of trajectories Descriptive chronology of the activity
Non-instrumented observations of activity using observation grids/self-monitoring grids (over 14 months)	Developmental trajectory analysis Context analysis	Formalization of trajectories Descriptive chronology of the activity
Instrumented observations of the activity by video and audio recording: cameras and dictaphones	Analysis of verbal and non-verbal interactions	Transcription Verbal/non-verbal interaction analysis grid
Self-confrontation interviews on video extracts	Transcription, corpus analysis	Characterization of the teams Analysis of the action and the situated acceptance
Group interviews with interview guides	Transcript Text corpus processing with QDA Miner software	Thematic analysis by major project stage Characterization of teams
Collecting the traces of the activity: e-mails, progress reports, project presentation slides, working documents, etc.	Description of the construction of the use and development of the traces	Developmental trajectory analysis Analysis of situated processes

Table 10.1. *Methodological protocol*

10.4. Results

In the Etui-Innovant project, we positioned the development of the design trajectory and the development trajectory of the innovation object equally, in order to determine the role of the innovation object in the team's cooperation.

The objective of the Etui-Innovant project was to design a waterproof and indestructible case (or, in French, *étui*). The development trajectory of this project was coupled by the construction trajectory of this case. We present in parallel the development of three processes: collective intentions, collective actions and the formalization of the innovation object.

Step 1: Starting point of the project

This first table (Table 10.2) on the birth of the project indicates the nature of the collective intentions, actions and innovation object. The bidirectional arrows in the table indicate that these factors exerted a mutual influence. Indeed, while the initial intention was for the team to evaluate its capacity to undertake action, this intention was realized by several research actions concerning ideas on the internet

Collective intentions	Collective actions	Formalization of the innovation object
Find an innovative concept to create a project ⟷	Research of problems and needs, definition of entrepreneurial objectives ⟷	Drafting of the company's values and the innovative part of the object: "Object enabling us to keep valuable paper records"

Table 10.2. *The starting point of the project (Step 1)*

Step 2: Integration of the entrepreneurial class and redefinition of the project

In this second stage, the schematization of the innovation object on paper and then in a digital format allowed the team to materialize the idea and validate it with a scientific team. The technological object reflected the values and ideas of the entrepreneurial collective.

Collective intentions	Collective actions	Formalization of the innovation object
To keep the original idea **To have the team validate the idea of the entrepreneurial class** ⟷	Market research, field survey ⟷	Schematization of the case (paper drawing, then digital drawing)

Table 10.3. *Redefining the project (Step 2)*

Step 3: Designing the digital mock-up of the case

In this stage, the innovation object went from the schematic to the model. More detailed and functional, the mock-up was a true revelation of the constraints of the project in terms of both the design and the initial idea. The innovation object then became a self-evaluation and correction tool (design and functions of the tool), as well as for the project, its trajectory and the objectives of the entrepreneurial team.

Collective intentions		Collective actions		Formalization of the innovation object
To evaluate the design constraints of the innovation object	↔	Digital mock-up of the case on specialized CATIA software	↔	Digital mock-up showing case dimensions and design materials

Table 10.4. *Design of the model (Step 3)*

Step 4: Design of the material model of the case

The innovation object took the form of a material model, a prototype of the object to be industrialized and then marketed. This innovation object acted as a boundary object between the entrepreneurial team, the educational and scientific teams, the industrial partners, etc. It was both an object of evaluation (and self-evaluation) and of communication around the project.

Collective intentions		Collective actions		Formalization of the innovation object
To obtain a life-size model and evaluate the resistance and watertightness of the model **To have a medium to promotethe project to funders**	↔	3D printing of the digital model, carrying out the various resistance tests with an industrialist	↔	Material model of the case 

Table 10.5. *Design of the model (Step 4). For a color version of this table, see www.iste.co.uk/bobillierchaumon/digital.zip*

The previous table (Table 10.5.) shows a change in the role and status of the innovation object, which changed from an intermediate object (Vinck 1999) to a boundary object (Vinck 2009), successively and cyclically. The development of the object's trajectory is linked to the progress stage of the project, the collective intentions of the team and the actions taken. Finally, the different processes that took place in this activity were interdependent and influenced each other. They needed to be studied together in order to understand the activity in all its depth.

The design processes described for the Etui-Innovant project and the Fauteuil-Innovant project demonstrate a close link between collective intentions and the formalization of the innovation object. This object corresponds to both the purpose and the result of the activity (Engeström 2009). The formalization state of the innovation object contributed to the construction of new intentions that potentially ran counter to the previous intentions. Thus, the intentions and the innovation object evolved together, through the actions implemented.

The innovation object occupied different roles during the activity. It could constitute an intermediate object (Vinck 1999) when it was used within the project team, particularly in design. It became a boundary object (Vinck 2009) when its formalization allowed it to be understood by actors outside the project and enabled a dialogue with external actors. Finally, the innovation object simultaneously fulfilled intermediate and boundary functions. In this case, the object, which was still being designed, and therefore modified within the project, was still used as a medium for communication and negotiation by team members. Figure 10.1 shows the different changes in the status of the innovation object through the development of the project.

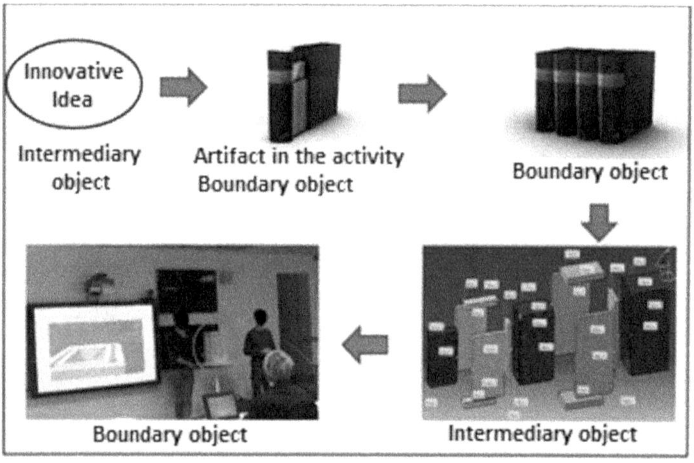

Figure 10.1. *Evolution of the format and roles of the innovation object in the Etui-Innovant project. For a color version of this figure, see www.iste.co.uk/bobillierchaumon/digital.zip*

It can be concluded that the traceability of the development of the project was partly dependent on the development of the innovation object. The trajectory of the entrepreneurial project took shape through the development of the innovation object. The representation of the innovation object, whether graphic or not, virtual or material, reflected the collective intentions of the members of the project team. It acted as an overview of their intentions at a given moment of the project, like a photograph of the project at a given moment of its progress.

10.5. Discussion and conclusion

The innovation object acted as a projection or a representation of the entrepreneurial intention that was common to all team members. This early representation of the innovation object was the basis on which the team delimited and structured its activity. The innovation object became an intermediate object that highlighted the material constraints of the project and served as a mediator and a time marker of the development of the objectives, and, therefore, of the trajectories. It served as a socio-technical framework for action in a cooperative work situation. The innovation object became a boundary object from the moment it was drafted and then modeled. These two stages enabled the development of the power to act for actors who were external to the project.

These analyses allow us to conclude that the selection of artifacts *from* the entrepreneurial activity was a crucial step in the development of the project. When selecting an artifact, stakeholders validated the artifacts' patterns of use[1] and mobilized these artifacts in subsequent situations. We will speak of a collective appropriation of the artifact, when it contributes to the cooperation within the team and the development of the project. This validation of the artifact was not done *a priori* but throughout the project. We therefore speak of a *situated acceptance* (Bobillier Chaumon 2013) of the device, which then became an innovation object for the team. The dimensions of the situated acceptance determined the longevity of the artifacts in the activity as follows.

The organizational dimension refers to the role of artifacts in organizational emergence, the prioritization of functions and tasks, the negotiation, decision-making and planning. Entrepreneurial teams only retain the artifacts that were resistant to complex work organizations and that adapted to organizational change. Hybrid (formal/informal) environments were particularly valued by teams, as were collaborative platforms that allowed for real-time discussions. These devices were retained and mobilized throughout the projects in a variable way.

The relational or interpersonal dimension was essential for the validation of artifacts and collective work environments. They needed to be "facilitators" in the interactions between actors. Artifacts that hindered interactions were quickly

1 Or social patterns of use (Rabardel 1995), which are models of how an artifact is used in collective activities. These social patterns of use are created in real-life situations of use.

rejected by teams because they disorganized the collective (Bobillier Chaumon 2016).

The professional and identity dimension was also very decisive in the longevity of the technological artifact. The results showed the role of the artifact in the identity and ideological construction of teams. The artifact needed to reflect the values, rules and know-how of the players. This dimension refers to the symbolic properties of the object. The players sought to recognize themselves in the instruments of their activity. We note, in our analyses, that the innovative and technological nature of the artifacts were a factor in their acceptance for teams presenting themselves as innovative. This last observation seems to explain the need for these entrepreneurial teams to develop or even change the objects of their activity on a regular basis.

10.6. References

Bobillier Chaumon, M.E. (2013). Conditions d'usage et facteurs d'acceptation des technologies : questions et perspectives pour la psychologie du travail. HDR Thesis, Université Pierre Mendès-France, Grenoble.

Bobillier Chaumon, M.E. (2016). Acceptation située des TIC dans et par l'activité : premiers étayages pour une clinique de l'usage. *Psychologie du travail et des organisations*, 22(1), 4–21.

Bonnardel, N. (2012). Designing future products: What difficulties do designers encounter and how can their creative process be supported? *Work, A Journal of Prevention, Assessment and Rehabilitation*, 41, 5296–5303.

Bouchard, V. and Bos, C. (2006). Dispositifs intrapreneuriaux et créativité organisationnelle : une conception tronquée ? *Revue française de gestion*, 161(2), 95–109.

Bowker, G. and Star, S.L. (1999). *Sorting Things Out: Classification and Its Consequences.* MIT Press, Cambridge, USA.

Buisine, S. (2013). Ergonomie pour l'innovation. HDR Thesis, Université Paris Descartes, Paris.

Engeström, Y. (2009). The future of activity theory: A rough draft. In *Learning and Expanding with Activity Theory*, Sannino, A., Daniels, H., and Gutiérrez, K.D. (eds). Cambridge University Press, Cambridge.

Folcher, V. and Rabardel, P. (2004). Hommes-artefacts-activités : perspective instrumentale. In *L'ergonomie*, Falzon, P. (ed.). PUF, Paris.

Ianeva, M., Ciobanu, R., and Vacherand-Revel, J. (2017). De l'aménagement des bureaux à la transformation des modes de travail : quelles répercussions collectives ? *Psychologie du travail et des organisations*, 23(2), 137–158.

Lavoisy, O. and Vinck, D. (2000). Le dessin comme objet intermédiaire de l'entreprise. *Communications organisationnelles. Objets, pratiques et dispositifs*, Presses Universitaires de Rennes, 47–63.

Poidi, I., Vacherand-Revel, J., and Guibourdenche, J. (2016). La place d'un dispositive socio-technique dans la trajectoire de formation d'ingénieurs au travail coopératif de projet. In *Actes du 18ème congrès international de psychologie du travail et des organisations*, Salengros, P., Di Fabio, A., Lemoine, C., and Majer, V. (eds). L'Harmattan, Paris.

Rabardel, P. (1995). *Agir avec des instruments : de l'outil aux systèmes techniques, une approche cognitive*. Armand Colin, Paris.

Safin, S. (2011). Processus d'externalisation graphique dans les activités cognitives complexes : le cas de l'esquisse numérique en conception architecturale individuelle et collective. PhD Thesis, Université de Liège, Liège.

Star, S.L. (2010). Ceci n'est pas un objet-frontière : réflexions sur l'origine d'un concept. *Revue d'anthropologie des connaissances*, 4(1), 18–35.

Star, S.L. and Griesemer, J. (1989). Institutional ecology, "Translations" and boundary objects: Amateurs and professionals in Berkeley's Museum of Vertebrate Zoology. *Social Studies of Science*, 19(3), 387–420.

Vacherand-Revel, J. (2017). Le travail coopératif d'équipes de projet d'ingénierie à l'épreuve de l'activité en réunion médiatisée et multi-localisée. *Psychologie du travail et des organisations*, 23(2), 89–116.

Vinck, D. (1999). Les objets intermédiaires dans les réseaux de coopération scientifique. Contribution à la prise en compte des objets dans les dynamiques sociales. *Revue française de sociologie*, XL(2), 385–414.

Vinck, D. (2009). De l'objet intermédiaire à l'objet-frontière : vers la prise en compte du travail d'équipement. *Revue d'anthropologie des connaissances*, 3(1), 51–72.

Zarifian, P. (1996). *Travail et communication. essai sociologique sur le travail dans la grande entreprise industrielle*. PUF, Paris.

PART 3

Psychosocial and Socio-organizational Impacts of the Diffusion of Technology

11

The New Physical Territories of Digital Activity

This chapter questions the outlines and issues related to space as an object of research and action for work psychologists. Many companies choose to deploy flexible workspace solutions that involve non-assigned desks. Our contribution seeks to address the psychological and psychosocial impacts of these complex social and technical arrangements. In order to do so, we draw on research carried out within the IT subsidiary of a large French group and present different theoretical models likely to account for the transformations of space, as embedded within activities.

11.1. Introduction

The spread of digital tools is considerably reshaping the spatio-temporal boundaries of work (Valenduc and Vendramin 2016) and is contributing to the emergence of a new organizational model that values autonomy, trust and communication. In today's organizations, employees are encouraged to become actors in the transformation of their work practices and skills development (Ughetto 2007). In this context, many tertiary companies choose to deploy flexible workspace solutions, that is, shared offices, unallocated desks and open spaces. From a managerial point of view, the implementation of these office solutions are justified, among other things, by the indirect benefits (Evette and Fenker 2011) that they can bring, particularly in terms of individual and collective performance or knowledge sharing. In other words, the workspace tends to become an object of

Chapter written by Maria IANEVA, Raluca CIOBANU and Chiara LAI.

management, endowed with new functions, in a dual sense of the term, that of an instrument to control and organize work (Benedetto-Meyer and Cihuelo, 2016; Pillon 2016) and that of a device requiring specific management.

Implemented by major IT groups such as Microsoft and Google, among others, these workspace solutions are associated with a vision of work (Humphrey 2014) that is free from the constraints of space and time and traditional forms of organization, based on control by the presence and visibility of workers (Felstead *et al.* 2003 cited by Taskin 2010). The flexible or activity-based workspace thus embodies an ambition for managerial innovation and is a tangible extension of the New Forms of Work Organization (NFWO[1]) (Ajzen *et al.* 2015).

Research in the field of management has sought to understand the "phenomena that accompany these transformations in the physical work environment" (Maclouf 2011). Most of this research focuses on the furniture, or the ambient physical characteristics of the workplace environment, considered as a backdrop for the subjects' activity. The aim of these studies is to question performance or job satisfaction (Chabault and Martineau 2018). However, these office solutions are in fact complex material, social and technical devices whose psychological and psychosocial impacts are not sufficiently addressed by work and organizational psychology. In fact, few works in psychology (and particularly in work and organizational psychology) address these issues and teach about space as a research object in its own right.

Thus, our contribution questions the outlines and issues related to space as an object of research and action for work psychologists. In order to do so, we must first expand on the links between the transformation of workspaces and the transformation of work and employment, in order to highlight the articulations between organization and space. In particular, we seek to characterize the underlying model of the subject and, as a corollary, the model of activity, which guide the design and organization of these new flexible workspaces in order to highlight their psychological and social implications. In order to support this analysis, we will report on research carried out within the IT subsidiary of a major French group, the results of which have been presented elsewhere (see Ianeva *et al.* 2017). How do these socio-technical configurations contribute to establish and sustain a specific relationship with work on the intra-psychological, or subjective, as well as the psychosocial level?

---

1 Terminology may vary: non-territorial space, flex-office, activity-based environment, combi-office, etc. All of these terms refer to shared, unallocated workstation solutions, offering a variety of workspaces with predefined functions.

Secondly, we will engage in a reflection on the theoretical models likely to account for space and its transformations. We have chosen to present work from the fields of situated action and cognition (Lave 1988), environmental psychology (Fischer 1990), as well as cultural-historical approaches to activity (Vygotsky 1997; Clot 1999; Engeström 1999). Each of these approaches provide means of constructing space as an object of research and action.

Finally, in the conclusion of this chapter, we return to the intervention of the work psychologist in the context of the transformation of workspaces.

11.2. Transformation of spaces and transformation of work and employment: "spatialized work"

Paid work brings workers into a space separate from the domestic sphere, formally dedicated to productive activity. The delimitation (physical or geographical) of the productive space thus contributes to organizing, technically and legally, the worker's entry into a redefined regime of obligations, that of paid employment (Pillon and Vatin 2007). In other words, the space dedicated to production contributes to organizing the wage relationship and to constructing productive activity as an "abstract entity, independent of the content of work" (Billiard 1993).

As such, workplaces are part of a more general movement of "economic empowerment" (ibid.) which, in our view, considerably reconfigures both the social relationship and the worker's subjective experience. We readily agree with Billiard (1993) when, from what she describes as a socio-historical perspective, she states that the relationship with work is also affected "by the transformations of everything that escapes the analysis of work as a concrete activity", but which "nevertheless contributes to the definition of the status of work and the worker's place in society"[2] (p. 32).

The idea of the productive space as a medium for a particular type of social relationship, as a resource for the organization of Taylorian work, leads us to put forward the hypothesis that the current reconfigurations of work spaces, supported by the development of information and communication technologies (ICT), refer,

2 Such as, for example: "employment status, type of contract, method of remuneration, benefits, promotion, social benefits and protection, training, union rights ..." (Billiard 1993, p. 32).

among other things, to a transformation of the employment relationship[3]. These spatial transformations therefore cannot be understood and analyzed outside the organizational, social or societal context in which they take place. The challenge of such an approach is indeed the psychological analysis of the activity.

11.3. From "spatialized work" to the division between space and work

The productive space, whether it is a factory or an office, has long been considered an "arena" where production is deployed and organized. The Taylorian, and later Fordist, factories incorporated a mechanistic layout logic, with the aim of optimizing employees' time and energy, as well as facilitating control over their work (Lautier 1999). As Fischer (1978) points out, "the firm organizes space like it organizes time" (p. 400). This mechanistic design of workplaces gives space an infrastructure status, that of a technical aspect of production. Thus, its layout takes account of so-called "non-thinking" elements (Lautier 1999) such as the movement of workers, or the movement of materials and equipment, which can be optimized with a view of the overall functional rationalization of the production effort.

During the first half of the 20th Century, the principles of office organization borrowed largely from the industrial world. Since the 1960's, the development of tertiary spaces has gradually become a medium and object of managerial action (Cihuelo 2016; Pillon 2016[4]), to the point where it is now invested in with the power of performative action in regards to the occupants and their activities.

According to Pillon (2016), the spread of the office landscape model (Bürolandschaft) marked the first movement of a real division in the joint organization of spaces and activities. This is reflected in the succession of different layout models (action office, combi-office, flex-office) closely linked to managerial discourse on work efficiency. In concrete terms, this means a new economy of spaces, now "invested in with managerial representations and organizational aims on the conduct of work" (Cihuelo 2016, p. 58). Thus, the office landscape appears to be a response to the traditional offices' partitioning; the reconfigurable office of the

3 The hypothesis of the links between the transformation of spaces and the transformation of the employee–employer relationship was put forward by Mr. Fenker in several oral communications to the CNAM in 2017 and 2019.

4 Several authors review the history of spatial models (Fischer 1989; Pélégrin-Grenel 1994; Lautier 1999; Schronen 2003). We do not present an exhaustive review of these works, but we have chosen to bring together those which, in our view, problematize the joint transformations between space and work organization and highlight their specific features.

1970's (known as the action office) is based on a theorization of both work and human needs (Saval 2014; Pillon 2016) and aims to provide a spatial "solution".

It is the development of flexible forms of management that are nowadays promoted in managerial discourse, highlighting their beneficial, even "magical", effects on productivity, cross-functionality and employee well-being (Jemine 2016). Office design is thought of as an instrument, which enables the organization to adjust to the strategic and operational uncertainties that are inherent in the instability of the economic environment (Evette and Fenker 2009). Emphasis is placed on the "transformative" potential of space, which underpins the development of organizational structures towards greater responsiveness to market uncertainties (Ianeva *et al.* 2017). Office design is thus presented as having an immediate, unmediated impact on employees and their activities (Vandersmissen 2015).

We note that the Taylorian organization, associated with an exact and precise prescription of the work (station), is replaced by a moving and flexible organization, whose objects and means of prescription have evolved considerably. The "infrastructure-space" of the Taylorian workshop[5] is invested in with new functions. As a medium for a practice of managing an implementation of work, it becomes the instrument of the organization that seeks to discipline thought, that is, to "equip" the choices of a worker, who is now mobile, autonomous and struggling with an unpredictable professional activity. In other words, while the productive space has made it possible to consolidate a new type of social relationship – the wage-labor market – and to support the intensification of mass production, it is today the bearer of a (normative) conceptualization of what work activity should be. The challenge is less to contain production and the human resources associated with it, than to support the "distribution" of this activity in multiple spaces-times ("anywhere", "anytime" work), that is, the division between space and work.

11.4. Flexible work environments: from work to "activity"

As noted in the introduction of this chapter, a growing number of companies are choosing to deploy flexible (or "flex-office"), "self-service" or "activity-based" work environments. Behind the multiple names, we find some common organizational principles that will help us to better understand these arrangements. These include (1) the pooling of all resources, from the waste bin to the printer; (2) the sharing of workstations (non-assigned desks), more or less extended depending on the configuration; as well as (3) the range of spaces (and workstations) with

5 The open space of the tertiary office, so much decried in press and literature, is a direct extension of this. We find the idea of surveillance through the presence and visibility of workers in a decompartmentalized and shared space.

predefined function(s) and modalities of use (spaces for creativity, dedicated to collaboration, concentration, etc.).

11.4.1. *The example of the design of the workspaces of a large company: the reconfiguration of work areas*

Thus, when the IT subsidiary of a large group decided to rethink its environment and working methods, with a view, among other things, on promoting inter-business knowledge and "inter-team agility", the choice of the company's management was that of the so-called "activity-based" working environment. An ambitious project of regrouping more than 1,500 employees, previously spread over four sites, was under way. The preferred solution involved around 1,200 workstations shared over four floors in a building in the suburbs of Paris. The objective was to "simplify and reduce the floors"[6] by setting up a single circulation system, grouping together user-friendly spaces and decompartmentalizing the offices. It aimed to "strengthen the reception and cross-functionality" and needed to "guarantee flexibility of use according to the professions".

The standard layout of the four floors that were refurbished included three conviviality zones (an "*Accueil café*" (welcome café), "*Travail café*" (working café), "*Création*" (creation space)), reservable meeting rooms, "bubbles", that is, non-reservable closed offices, open spaces and open spaces with a variety of work stations. An "address" or reference area was assigned to each team. Teammates were invited to move in a privileged manner while respecting the "clear desk" policy and the principle of desk sharing. The team zones that were needed therefore provided reference points for users, without compromising the overall functioning of the spaces. The new office was thus presented as a complex arrangement of spaces and resources shared on a "self-service" basis, for which access and the terms of use were conditioned by formal and explicit rules. Let us take a few examples, based on the presentation document distributed to all employees upon their arrival at the new site – the "welcome booklet".

"Bubbles" were closed, non-reservable offices, which helped with "accommodating the needs of concentration and exchange", avoiding noise in work spaces and/or to maintain confidentiality, "Maximum recommended duration of use: two consecutive hours". The "exchange station" was intended for collaborative work with colleagues: "I set up here to work because I am ready to exchange or be interrupted". The "semi-concentration" office welcomed users working "alone or in

---

6 Project document extracts.

interaction" and performing "tasks requiring intermediate concentration". In contrast, the "concentration" workstation allowed users to "work alone" and perform "tasks requiring concentration".

11.4.2. *From space allocation to the redefinition of associated tasks*

We note that each type of workspace or desk was explicitly associated with a range of tasks and recommended behaviors (telephoning, exchanging, collaborating, brainstorming, working alone, concentrating), as well as specific modalities of use (settling down for a few moments, not exceeding two consecutive hours of use). The criteria for choosing a workstation were also specified in the "welcome booklet", as well as on posters presented directly on the workstations or near the spaces, or at various "live" events (workshops, coffee breaks, etc.).

The implemented layout solution materialized an "interpretation" of the work of the professionals. The rules specified were related to the general characteristics of the work broken down into "generic activities" (interaction, concentration, management, etc.) which justified the principles and modalities of occupation of the office (job sharing, duration of use, etc.). As a corollary, the material arrangement we have just described incorporated and participated, from the outset, in the organization of a conflict situation: the systematic failure of employees to comply with the operating principles and rules of use could potentially put the overall functioning of the activity-based work environment being tested at risk. It should be recalled that the office solution provided fewer stations than people (the ratio used was eight stations for every 10 people), as well as mainly open-plan offices.

As we pointed out earlier, the activity-based office we have described was intended to "organize", that is, to have a direct action on the behavior of users. We can therefore hypothesize that its deployment aimed to organize new modalities of action on work through relation and by not acting on (the content of) the task itself, as could be the case with the prescription of work at the time of the Taylor system, but by configuring a specific relation of/to work.

On one hand, it was a matter of establishing a shift between "work", "relations at work" and "interpersonal relations". By relying on generic categories of activities, the "design" of such solutions provided a framework for interpretation, that is, thinking and acting, decorrelated from the concrete reality of work. Paradoxically, by emphasizing interpersonal communication, supported by physical proximity in space, as a resource for efficiency and innovation at work, these socio-technical devices screened off the work being done. This raises the question for the occupational psychologist – intervening in the workplace – of how to (re)introduce work as an object of dialogue and thought in a context, where one seeks to "problematize" work issues in an interpersonal and cognitive register.

On the other hand, the users of activity-based offices were collectively given a new responsibility to ensure the proper functioning of their work environment.

Finally, in an indirect way, the design of this workspace solution exhibited, or even sought to affirm, a presupposition relating to the functioning of the subject, based on a model of a voluntary and rational individual-subject, whose intentions could be translated into acts. Moreover, change management in these real-estate projects were organized around employee experiences, considered in terms of adaptation to novelty, which opened up a strategy based on an interpretation of behaviors as "departing from the rule". By giving a normative orientation to this premise, what we seek to provide through the rules of use set out above is indeed the deliberation *in situ*, of the choices to be made in the context of a "self-service" work environment, where multi-activity and task fragmentation are developing increasingly (Licoppe 2007).

11.5. What theoretical models for considering space and its transformations?

Generally speaking, the workspace remains an under-investigated object of study in the humanities and social sciences. In the field of psychology, environmental psychology (Fischer 1989; Vischer and Fischer 2005) tends to consider space as a research object in its own right. Research in environmental psychology focuses on the modalities and individual forms of appropriation of workspaces through the notions of "territoriality", "appropriation" and "comfort" at work. In this perspective, offices and their layout are thought to be both socially organized and organizing. Workspaces embody the hierarchical system and participate in its operation (control of the productive process, access management, circulation of information) (Fischer 1990). This line of research has the merit of constituting offices as a research object in their own right. However, while recognizing the social character of workspaces, environmental psychology equates them with a "milieu", which is certainly organized, but whose recomposition by the individual is considered in generic terms, that is, with little connection to the context of productive activity and its development. Thus, individual appropriation behaviors reflect "a fundamental tendency of the human being" (Fischer 1978, cited by Cihuelo 2016, p. 60) and pose identity-related challenges.

Jean Lave's work, from the perspective of situated cognition, makes it possible to reaffirm the social and material character of the environment. Beyond a static version of the environment as an embodiment or product of social activities, it finds itself re-inscribed in the dynamics of both a concrete practice, in the process of being made, and a structuring arrangement, caught in an extended temporality. In his book "Cognition in practice. Mind, mathematics and culture in everyday life" (1988), Lave offers a dialectical analysis of the subject grappling with their environment. The distinction between arena and setting constitutes a form of operationalization of this principle of analysis. Indeed, she proposes to distinguish, on one hand, that which is part of a "sustainable entity", what is "already there" before the subject's action – what she calls the *arena* – from the environment taken over, reordered by and in the subject's action – the setting. Dialectical analysis takes the relationship between subject and environment as its starting point, which it institutes as mutually constitutive (rather than caught in a cause and effect relationship). In this sense, the environment is a relative term in that, on one hand, it is always related to a subject in action. On the other hand, it is never finished but always in the process of being done. As such, we could put forward the idea that the setting (as much as the arena) constitutes a developing process and is a conceptual contribution that is particularly convergent with Russian cultural-historical psychology (Vygotsky 1994, 1997).

Activity theory (one may say "activity theories") brings together a set of approaches, in the continuity of L.S. Vygotsky's seminal work, that are in many ways, "heterogeneous". They are, however, articulated around a certain number of common proposals of the organizing principles of action. The unit of analysis selected by these approaches is that of the activity, or system of activity, according to the terms of Engeström (1999), oriented towards an object. Collective activity, thought of as a dynamic and distributed process, constitutes the context in which the actors' situated practices take place. The aim of the activity refers, on one hand, to the concerns shared by the collective, generating attentional focus and meaning. On the other hand, the object corresponds to the path, the trajectory towards the product of the activity (Engeström 1999). Through the mediation of artifacts, rules, the division of labor and the community to which the actors belong, this horizon of potentialities draws "salient configurations" for cognition and action (Ianeva 2012) and guides without ever determining.

The principle of double stimulation (Vygotsky 1997) is an interesting entry point for highlighting the advantages of these approaches, for working on transformations in workspaces and their psychological and social impact. This principle could be analyzed as a general mechanism that allows subjects, caught in a conflict situation, to intentionally change their practical and psychological conditions. As an illustration, Vygotsky (1997) used the example of a subject who is asked to participate in an experiment, only to be left in a separate room without explanation, thus creating the conditions for a conflict of motives – to leave or to continue waiting. When he chooses to rely on an external artifact (his watch) and to leave at the moment the hands of his watch reach a particular position, the conflict is transformed, that is, redefined in a way that affects the subject's reinterpretation of the overall situation. He is no longer grappling with a conflict of motives but has himself, rethought the terms of the problem with the help of an auxiliary stimulus (the watch). Analogously, the flexible work environment, by virtue of its organization and operating principles, helps to install a new conflictuality between the demands of the activity being carried out and those of the "collectivized" management of spaces, the reinterpretation or psychological reinvestment which would require the setting up of a specific framework of action. In a sense, we could argue that transformations of workspaces are opportunities to develop or, on the contrary, impede on the activity of professionals.

11.6. Conclusion

We have sought to question the outlines of current transformations in tertiary workspaces in order to outline the issues at stake for work and organizational psychologists. In order to do so, we returned to the tensions that characterize the relationship between space and work organization. While space has always been a medium for managerial action, it is now being given new functions, namely that of an "autonomous territory" (Ficher 2011) which has become a specific management object with direct and immediate action on workers' practices and efficiency. Workspace design is thus presented as a composite arrangement (furniture, activity-based areas, services) whose aim is to equip the choices and thinking of an autonomous, committed and responsible subject. As such, the current transformations of workspaces can be understood as part of a wider movement towards the individualization of work.

In this context, we have particularly sought to highlight how flexible workplace solutions embody a (re)interpretation of work in terms of generic "activities" (concentration, collaboration, innovation, etc.) and also give them a normative orientation. The risk is then, as Evette and Fenker (2011) point out, to contribute to organizing and thus "perpetuating a situation of permanent change" at work with deleterious consequences for workers' health. We will add that experimenting activity-based workspaces entrust users with collective responsibility for the quality of their working environment. This leads to a new demand for psychosocial regulation of individual and collective behavior, often unthought of by companies and therefore insufficiently recognized and instrumented. This dual requirement – to carry out one's own activity in the best possible conditions and to participate in the collective management of spaces – can, in our opinion, be as much a hindrance as a resource for occupational health. The challenge for the work psychologist intervening in the workplace would then be, based on a detailed and lucid analysis of these transformations, to use them as a lever to organize the conditions of his or her own intervention.

11.7. References

Ajzen, M., Donis, C., and Taskin, L. (2015). Kaléidoscope des nouvelles formes d'organisation du travail : l'instrumentalisation stupide d'un idéal démocratique et collaboratif. *Gestion 2000 : management and prospective*, 31, 125–148.

Benedetto-Meyer, M. and Cihuelo, J. (2016). L'espace dans l'analyse du travail. Présentation du corpus. *La nouvelle revue du travail*, 9.

Chabault, D. and Martineau, R (2018). Les formes de contrôle dans les organisations multi spatialisées : post-bureaucratie ou persistance de la bureaucratie ? *Actes du XXVIIème Conférence Internationale de Management Stratégique*, Montpellier, 6–8 June.

Cihuelo, J. (2016). Les dimensions cachées du travail en open space : le cas de téléconseillers en centres d'appels. *Sociologie du travail*, 58(1), 57–79.

Clot, Y. (1999). *La fonction psychologique du travail*. Presses Universitaires de France, Paris.

Engeström, Y. (1999). Communication, discourse and activity. *The Communication Review*, 3(1–2), 165–185.

Evette, T. and Fenker, M. (2011). Conception et usage de l'espace – une autre lecture des risques psychosociaux. In *Risques psychosociaux, quelle réalité, quels enjeux pour le travail ?* Hubault, F. (ed.), 167–177. Octarès, Toulouse.

Fischer, G.-N. (1978). L'espace comme nouvelle lecture du travail. *Sociologie du travail*, 20(4), 397–422.

Fischer, G.-N. (1990). Espace, identité et organisation. In *L'individu dans l'organisation. les dimensions oubliées*, Chanlat J.F. (ed.), 165–183.

Fischer, G.-N. (2011). *Psychologie sociale de l'environnement*, 2nd edition. Dunod, Paris.

Humphry, J. (2014). Visualising the future of work: Myth, media and mobilities. *Media, Culture and Society*, 36(3), 351–366.

Ianeva, M. (2012). La restructuration du travail chez Santé Info : du développement de l'activité d'un centre de contact spécialisé à ses incidences sur les pratiques situées. PhD Thesis, Université Lyon 2, Lyon.

Ianeva, M., Ciobanu, R., and Vacherand-Revel, J. (2017). De l'aménagement des bureaux à la transformation des modes de travail : quelles répercussions collectives ? *Psychologie du travail et des organisations*, 23(2), 137–158.

Lautier, F. (1999). *Ergotopiques : sur les espaces des lieux de travail*. Octarès, Toulouse.

Licoppe, C. (2007). De la communication interpersonnelle aux communautés épistémiques : le développement des TIC et l'enracinement du paradigme de la distribution. *Hermès*, 47, 59–67.

Maclouf, E. (2011). Espaces de travail et management. *Revue de gestion des ressources humaines*, 3(81), 5–18.

Pillon, T. (2016). Retour sur quelques modèles d'organisation des bureaux de 1945 à aujourd'hui. *La nouvelle revue du travail*, 9.

Pillon, T. and Vatin, F. (2007). Retour sur la question salariale : actualité d'un vieux problème. In *Le salariat : théorie, histoire et formes*, Vatin F. and Bernard S. (eds), La Dispute, Paris.

Saval, N. (2014). *Cubed: A Secret History of the Workplace.* Doubleday, New York.

Taskin, L. (2010). La déspatialisation. Enjeu de gestion. *Revue française de gestion*, 3(202), 61–76.

Ughetto, P. (2007). *Faire face aux exigences du travail contemporain. conditions de travail et management.* Éditions de l'ANACT, Lyon.

Valenduc, G. and Vendramin, P. (2016). *Le travail dans l'économie digitale : continuités et ruptures*. Éditions ETUI, Brussels.

Vischer, J.C. and Fischer, G.N. (2005). User evaluation of the work environment: A diagnostic approach. *Le travail humain*, 68(1), 73–96.

Vygotsky, L.S. (1994). The problem of the environment. In *The Vygotsky Reader*, Van der Veer R. and Valsiner J. (eds). Basil Blackwell Ltd, Oxford.

Vygotsky, L.S. (1997). The history of development of higher mental functions, Chapter 12: Self-control. In *The Collected Works of L.S. Vygotsky, The History of the Development of Higher Mental Functions*, Rieber R.W. (ed.). Plenum, New York.

12

Digital Work, Disposable Work? When Opportunities to Explore Threaten the Meaning of the Activity

Attentive to the motives for commitment in work, this chapter points out the ambivalence of video game workers towards their digital, creative and collective work. After having exposed the mode of division of this complex digital work, we show the possibilities of exploration and management of its contingencies offered to the employees by the chronic indeterminacy of the manufactured product. However, we will also see that its digital nature, by allowing multiple iterations, also feeds the fear of disposable work and may weaken their commitment to the work.

12.1. Introduction

Digital technology is often presented as an autonomous, abstract and unrelenting process, threatening to destroy jobs and/or working communities. Consequently, it is evaluated in binary terms, as a factor of either the emancipation or alienation of workers (Bidet and Porta 2016). This naïve image is commonly associated with the idea of the substitution (with jobs or products) of a robot for one or more workers, and so hardly encourages us to look further into the concrete technicality of digital technology or digital work. Yet, for the most part, the digitization of goods and services does not mean a replacement but a transformation of the technical nature of human work, which is more widely mediatized: increasingly, it is a matter of acting *with* digital artifacts, that is, *through* "reading and interpreting signs" or "signals of signals" (Naville 1963). Naville has been working since the 1950s on the

---

Chapter written by Maxime BESENVAL and Alexandra BIDET.

"restless psychology" of the controllers of these automated devices, "moved and moving with almost no human intervention". Since their activity is more a matter of monitoring or investigation, they may feel that they have lost "direct control" over the production process and no longer act "on their own". According to Naville's formulation, the archetypal figure of human work is then no longer the lifting of loads – according to the mechanistic conception of work as an expenditure of motor force – but of coping with contingencies. In these contexts, activity is in fact widely distributed between humans and digital devices – leaving the former responsible for intervening in the event of a drift or the whim of the flow, so plasticity or "proper human availability to change seems to be at the heart of the activity" (Bidet 2011a, p. 27).

However, digitization is also taking place in industries where human work, which is predominantly intellectual and symbolic, remains at the heart of the productive process. The digital nature of the product allows here another form of activity distribution: not so much between humans and artifacts as between humans themselves, between whom a digital product circulates that its materiality (it is made of symbols, "zero" and "one") makes continuously modifiable. The same problem thus arises from the dynamic nature of highly digitized work environments, where the product is a flow in perpetual transformation: how can we take hold of this flow? Is it still possible to identify a tangible personal contribution? While our most common images of work remain largely dependent on mechanistic conceptions (Vatin 1993), which associate it with an expenditure of motor physical force, a direct intervention on a material or on a productive process (Bidet 2017), the challenge of the development, made possible by digital technology, of mediatized and dynamic systems of activity is nothing less than another conception of what working means.

When activity slides towards interface work – bridging processes, activities and/or actors, it mobilizes humans less for their motor physical force, or even their intellectual capacities, than for their flexibility – their ability to face constantly changing and new situations or problems. The increasing share of contingencies can then make the final product of collective activity, and the effects of each individual's contribution, particularly uncertain. At the very least, it requires a particular investment on the part of each individual to continue to identify forms of "true work" in his or her activity, that is, activities that he or she believes are worthwhile (Bidet 2011a). Commitment to work can also be weakened in the long run by a sense of the work being disposable.

This chapter will thus ask how far we can invest in and develop the opportunities offered by digital transformations – in terms of the permanent and collective reconfiguration of the product of labor – without threatening the very possibility for workers to engage in it. To do this, we will rely on an ethnographic survey carried out on video game development within a pioneering company in the sector, which has about a hundred employees, where one of us conducted six weeks of observation and about 15 semi-directive interviews with employees of varied profiles. The activity of video game assemblers[1] will thus offer us the typical case of a system of activity where the digital nature of the product increases the possibilities of individual and collective exploration tenfold, while subjecting the work experience to a constant regime of contingencies. Subjected, rather than managed collectively, they then weaken its meaning and exercise: would this digital work also be, by its very nature, disposable work?

12.2. The division of complex digital work

The projects of the video game company we investigated span several years of development, with budgets reaching several tens of millions of euros. Each of these projects brought together a wide variety of expertise in the same workspace, with no spatial distinction between teams working in the same open space. Given the complexity of the product being developed (a complexity that takes into account both the technical challenge of releasing a game that complies with industry technological standards and the organizational challenge of coordinating numerous teams with multiple interdependencies) and the long timeframe for producing a game, development is broken down into intermediate phases: milestones are set by the publisher – which occurs every three months, with the publisher's payments being indexed to these quarterly deliveries – and internally by management every month. Longer-term project management is grounded in a third timeframe, determined by successive stages of development over periods ranging from 6 to 18 months, at the end of which the entire game is presented to the publisher, with an increasing number of criteria to be met. Here the pre-production, production, post-production and finalization stages follow one another. Each of these stages has its own quality criteria, which are centralized around three elements: the stability of the code, the certainty in the definition of the product, and the quantity and quality of the data integrated into the game. The pre-production phase is therefore a phase dedicated to prototyping, and the following are stages of production and refinement of the different elements of the game.

The nature of the work carried out is in fact extremely heterogeneous according to the various sequences. We find the model of the unique product (or site), outlined by Peter Drucker's typology, where the sequencing of activities responds to the

1 Here, we choose a generic name that does not have a direct correspondence in the profession, the modalities of division of labor and the names of the profession being specific to each company in the sector. Depending on the organization, they may be called level or junior game designer, scriptwriter, game builder, etc.

singularity of each project and sees a succession of differentiated productive phases. Beyond these temporal breakdowns, which allow the organization to set intermediate objectives for the employees and facilitate a common understanding of the progress of development, the project is also broken down in terms of its content, in order to set up cross-functional teams, each working on one part of the game (called "chapter" or "scene", according to a breakdown found in the final product). This model of organization in small teams, in charge of part of the game's content, also aims at a better "appropriation" of the scene on which the employees operate – a recurring theme in the discussions of the management staff.

The activity of the assemblers is marked by the central notion of "build", which is an update of the code and data, and by extension of the product, carried out three to four times a day. The entire production process thus consists of advancing the build (by integrating content, correcting identified problems, modifying parameters, improving graphics, stabilizing the code, etc.), until it reaches its final version, which will be distributed on the market. The characteristics of the product, if they are initially defined in a theoretical and abstract way (through the scenario, as well as descriptive documents of all the elements of the game), gradually gain independence from the textual and graphical material media, as the documentation can no longer follow the inflections given to the build by the multitude of micro-decisions made on a daily basis. If the watchword of pre-production is "*to refer to the documents, to get closer to the written vision*", it will be replaced little by little by a very different maxim: "*it is the build that is right*".

As each build represents a specific state of the product, updating the real knowledge of the project is a major challenge for the productive efficiency of the teams. This technological context is also essential for understanding the producers' activity. The role of updating the representation of the flow (Rot and Vatin 2017, p. 59) falls to them, both through formalized processes (a meeting with the other members of the production team takes place every day, and an activity report is sent to management at the end of each day and week) and informal relations (oral exchanges of information are omnipresent and give a strong discontinuity to the activity). It should be noted in this respect that other people's roles, such as the Quality Assurance Testers (whose main mission is to identify all the problems present on each build and to constantly update the game's bug database), rely on this ability to update and share knowledge of the state of the product. The injunction to "*test the build*" is a recurring theme that is tirelessly repeated by management and local managers. The speed at which the product is transformed is indeed very intense.

This technical environment introduces us to the challenges of the assemblers' activity, which is the main and most visible lever of the game's transformation. Being the indispensable intermediary for integrating the majority of the content developed by the other teams into the build, the activity of these operators is subject to both vertical expectations (coming from line managers through the objectives set) and horizontal expectations (the visibility of the work of other employees being dependent on the assembler's rate of progress). Assemblers are thus dependent on other teams to do their work, just as the rest of the company is dependent on them to make their work visible in the game. However, the relationships are rarely linear (according to a model that would see interactions limited to the disinterested supply of components, each one having no more control over its production once it is integrated into the final game); on the contrary, they are based on the creation of multiple ephemeral and fragmented spaces of collective work. Exchanges, back and forths, reciprocal explanations of technical specificities and common attempts to circumvent the respective constraints animate the workspace on a daily basis. It is therefore not a similarity of professions that creates proximity, but the moments of the activity that call for cooperative abilities, in exchanges and interactions:

> In my team, we don't have a lot of group consciousness. Part of it comes from the fact that our work [...] is quirky, because we have a job that's cross-departmental, where we interact with a lot of people who are not from ours. On the other hand, our work is very compartmentalized within our own team, so we have classic working relationships, because we're close geographically, but we don't work on the same things very much. We have few real reasons to talk about our work among ourselves, except to share knowledge (when someone has a specific technical problem, they come and ask the others if they have encountered the same problem). [...] But that's it, the only knot that binds my team is the work tool. (V., assembler)

Thus, relational forms emerge that are essentially based on "technical solidarity", which Nicolas Dodier defines as this "form of links between beings created by the functioning of technical assemblies" (1995, p. 15). Through the multiple situations in which constantly updated technical problems are confronted, the supposedly natural affiliations (here, the team of assemblers) thus lose their primacy. On the contrary, a community of condition emerges in relation to the product – in the face of its chronic indeterminacy.

12.3. Chronic indeterminacy of the product

Because the medium is interactive by nature, there is a shared understanding among video game production companies of the need for prototyping. Projections and theoretical specifications relating to the rules of the game, the ways of playing, and the content of the project are not enough to predict the future quality of the product. The game under development is thus inherently unstable and indeterminate until the teams have produced sufficiently complete versions of it (allowing a final decision to be made on the nature of the different elements of the game).

The challenge of the pre-production phase thus consists of validating all project components subject to uncertainty. This stage is present in various forms in all companies in this industry. The goal is to reduce the uncertainty affecting the fundamental characteristics of a project. By extension, the actors' references are always oriented towards "agile" methods. This conception of project management makes "adapting to change more than following a plan" a fundamental value. This translates into a cardinal operational principle: "welcome changes in needs, even late in the project"[2]. These principles are embodied on a daily basis in the interactions observable in the workspace. One of the creative directors, in order to justify a modification of an element of the game that has already undergone numerous inflections, thus signaled his willingness to standardize this process: "Changes are part of artistic experimentation, these hesitations are indispensable in an interactive medium such as ours" (meeting observation).

This ubiquity of prototyping, redefinition of software features, and back and forth between employees and creative management is not perceived negatively. Respondents consider this process to be an integral part of their job:

> Iterations are an important point, which allows us to refine the mechanics, to test them before embarking on their systematization. I would tend to say that this is vital. (Interview with A.)

> For me the iterative process is something really cool and interesting. And a successful iteration is: you put an idea down, you see if it works, and you can redo it from scratch if you see that it's catastrophic and it doesn't work at all. [...] You build a base, a direction to progressively move the product toward. (Interview with L., assembler)

Undoing in order to start over is thus an aspect of the activity considered as being positive, in particular because it involves technical exploration practices. Each prototype is a renewed opportunity to deploy a form of creativity in the face of a new problem; it commits to reconfiguring previous standards in a progressive hybridization of technical objects. Employees value these wide margins of singularization of the activity, in contexts that test their ability to deal with the unknown:

> What I'm interested in is when I can experiment. [...] Because it's not true that the technical implementation is a straight line, where you have to do A then B, and it works. There are plenty of ways to do it! [...] The design of the script that you can have when you're outside of

2 Manifesto for Agile Software Development, 2001.

> it, is an application, in the context of a decision, a design, which is outside of it. [...] We have a geometrical image, we tell ourselves that the most efficient way to join two points is a line. But in reality, with the bundle of constraints imposed by the external decision, it is impossible to find the perfect technical means for a given situation. That is to say, since the decision is bound to change, there are really many ways to apprehend what is asked of us and how to implement it. That's why I think it's non-linear, each assembler will interpret the decisions in his own way, and place the points of interest, let's say the logical points of the technique, in different places, to have a turnaround. (V., assembler)

It is no longer technical expertise that is the dominant quality here, but the ability, at once technical, relational and organizational, to adapt to a changing environment. Particularly valued is the ability to make the multiple interstices of "co-activity" (Rot 2017) fluid, and to quickly make mental projections from requirements/prescriptions still at the state of potentiality, so as to reduce uncertainty for the various stakeholders. Here, we find certain characteristics of the "state of greatness" in the project-based society (Boltanski and Chiapello 2011, pp. 182–187): being willing to change, capable of optimizing the temporal resource, versatile, gifted with an open and curious mind. These traits are no longer only valued in executives and managers, but also in subordinate categories, in addition to their technical skills. In this context, the multiplication of contingencies associated with digital work is not only the counterpart of opportunities for exploration. Coupled with versatile prescriptions, not managed collectively, it also threatens commitment to work.

12.4. When the contingencies regime threatens work commitment

While the primacy of contingencies may be associated with moments of exploration and creative projection, the prevalence of iterative development processes is also likely to threaten the very possibility of engaging in the activity. Indeed, the indeterminacy of the product brings a strong uncertainty to bear on the future of the work carried out, and thus the possibility of giving meaning to its experience. This uncertainty is all the more critical as it is accompanied by a feeling of opacity in the decision-making process, which is considered arbitrary, and a low visibility as to the temporality of the changes required and their impact on the final product.

The interviews yielded many accounts of situations, presented as traumatic, where several weeks of work were challenged by a creative change of decision. Several of our respondents, who have since left the company, highlighted prescriptive instability as the main reason for their departure. The elusive and unstable nature of tasks made it difficult to build activity continuity (Bidet 2011b) (which is then difficult to grasp as a process with a beginning and an end) and tangible consequences. This instability of prescription was a recurring topic of discussion among employees, in informal moments of discussion, during breaks or at lunch: "I don't even know what I did this week", "It feels like the scene is not moving, we change direction all the time", "We go backwards as soon as we start to move forward a bit, it's going to drive me crazy". The repetition of this type of situation makes it impossible to appropriate and singularize one's activity, which is a condition for being able to accomplish it (Bernoux 1981).

Bertrand, who has been with the company for more than ten years, bears witness to this. He told us that he has been worn down by unstable work and that he is too subject to the arbitrariness of creative decisions to be done "properly":

> Everything can change from one day to the next. It's very demotivating. And the worst thing is that when you have experience, you know it's going to be useless. It's not useful. It gets on my nerves. From the beginning I know when it's not going to help, but you have to do it. Making people work like idiots, on shit, without question, is to belittle them. It gets on everybody's nerves.

The experience of work is then no longer seen as an arena to respond to the imperatives of "technical solidarities" or to display one's virtuosity (Dodier 1995). It only responds to the need to protect oneself from a new experience of disregard for the work done:

> The thing is that I know the company so well that I don't want to bust my ass anymore. I wait until the last minute to do things, because that's when it's least likely to change. It's ridiculous, it's not a good way to work. That's where we're at.

All the respondents thus underlined the inability of creative management to make final decisions on the basis of prototypes.

During a discussion, several employees scoffed at this method of management:

> Ah no, but here creative work is good. We wait for everything to be polished, and then we decide what's killed or not. That way we don't get bored. We are officially told at meetings: "It should be polished so we can judge."

J. (producer) said :

> Yeah, we don't really like to ask ourselves the question before people have finished. You know, the famous question "Can we test it?" [he takes a theatrical look of disbelief, people around us start laughing]. We all know what testing entails. First, we test as a placeholder. Then they ask you to produce the data, integrate it, polish the gameplay. And then they'll make the decision. When you hear "can we test?", you know that your job can go in the trash, but only when it's done. Otherwise, it wouldn't be funny.

This devaluation of the activity is not specific to the company's subordinate jobs. It can also be seen in interviews with team leaders, who deplored these arbitrary and changing decisions, which they nevertheless had to relay:

> Let's just say you can take some work out of the garbage can by iterating once in a while. When it becomes systemic, on the other hand, it means it's adopted as a way of operating. And that's a problem. How do I justify it to my team? [...] It puts me in an impossible position. You can't be credible by giving deadlines or putting pressure on a team when it has months or even years of work behind it that has gone up in smoke. (B., team leader)

The criticism from the employees (assemblers as well as producers) was also rooted in an apprehension of the temporality of the activity. Prescription revisions were particularly criticized when they took place in advanced production stages (supposedly dedicated to refining product quality) or on elements of the game that had already been validated. The frustration caused by creative wandering in the late stages of projects was recurrent:

> I mean, we're doing something, it's not working, we're trying something completely different. It works at the beginning of the project, but after 4 years of development [...]. It's absurd what they ask us to do sometimes, to completely change scenes that are supposed to be "beta" [...] that have several years of development behind them [...]. It's completely crazy. (L., assembler)

The reaction also varied according to the legitimacy given to these requests for product modification:

> An iteration here is going to be "we're going to try this instead of this", but people don't bother, at least we builders don't bother, to justify why we're going to try B instead of A. We don't have any real

> justification. It bores a lot of people. […] For me, the iterations should be framed by an approach, which is clear, communicated to the team (especially those impacted by the modifications, because it's important that people understand why they made the modifications, how, in which direction we're going). It's entire support that is sorely lacking, which makes the iterations feel like something harmful […] in any case, annoying […] we know it's going to happen, we don't know how, we don't know what it's about, it seems to be totally by chance, it seems unmotivated. It's a big problem. […] Yet, it's one of the basics in video games, we're always told that we shouldn't be offended when we're asked to do the job again. But let's just say that it's so badly managed here that it's hard sometimes not to be offended by the fact that we're asked to take everything away. Because they can't explain it to us! Why do we have to? Why is it better if we do it? It should be explained why we do things, not just impose them, with the tone of it being obvious, with contempt. (V., assembler)

The academic resources of these video game professionals, as well as their previous professional experiences, are as much support for their criticisms of an unstable prescription as they are the direction of the requested changes. In this digital environment, the technical complexity of an element has little to do with its time of appearance on the screen or its importance in the eyes of the player. Consequently, repeated iterations on elements of the game that are technically very complex but with a minor visual impact accentuate the dynamics of rejection of the prescribed changes. After having spent almost three months working exclusively on an element appearing only a few seconds in the game, one of the project's assemblers expressed his dissatisfaction:

> I don't want to think that all I've done on this game is these two minutes! I can't go on like this. The time I spend on it, just because we're constantly changing everything […] frankly it's getting in knots in my brain. (M., assembler)

Also observed by Nicolas Auray in the case of IT developers (2003), the processes of "dissolving personal decisions into a network of collective rectifications", and of their "distillation into a complex network of connections", tend to undermine the possibility of identifying a significant contribution of one's activity to the final product.

The creation of video games is marked by a high degree of technological interdependence and the intervention of a multiplicity of expertise; a production model by stages, where the single product is continuously improved over a long period of time; the importance of interpersonal relationships and the absence of categorical labor management systems; an extreme instability of prescriptions; managerial systems that promote product ownership, open participation in decision-making, and a high degree of porosity between private and professional time; finally, a vocational commitment shared by employees who are passionate about video games. In this context, the commitment to work is all the more problematic as the injunctions to personalize the relationship to activity and expression are seldom realized, as iterative work poses a constant threat to the meaning of the work done, and as what employees value in their activity is often disqualified as unproductive or secondary by management. Multiple forms of resistance can be observed in this way. They may consist of modulating commitment, by refusing attendance standards, by using multi-activity to reappropriate time spent at work, or by displaying cynicism with regard to the values conveyed by the company's management teams. Resistance also involves attempts to transform the experience of work, by demanding greater formalization of prescriptions, promoting other performance criteria and/or modulating the care given to different aspects of work.

12.5. Conclusion

While the unforeseen can be an "essential opportunity for the development of creative processes" (Lhuilier 2016), its inclusion at the heart of the activity, permitted or increased tenfold by digital technologies, can therefore also, if not managed collectively, hinder engagement in the activity. Here, the absence of collective control of contingencies, which are therefore perceived by employees as resulting from arbitrary decisions, tends to hinder the dynamics of the singularization of the activity (Schwartz 2000), and therefore the employees' attachment to it, by causing chronic uncertainty about the usefulness of the work carried out. Under these conditions, the central place of contingencies threatens the meaning of the efforts undertaken, disrupts the decisions and support of the activity as well as the reference points of judgments (Bessy and Chateauraynaud 1995; Dodier 1995, p. 191). We can thus see how much the possibility of perceiving in one's activity "true work", that is, an activity worthy of commitment, depends not only on the content of the activity and its practical operations, but also on the potential usefulness that one can possibly associate with it: what is its value? This inquiry, which everyone conducts into the effects of his or her work, undoubtedly underpins the chronic democratic aspiration to "have a share in participating in determining the conditions and the aims of [one's] own work" (Dewey 1987, p. 233).

The fact that digital technology multiplies the possibilities of exploring and confronting contingencies, through revisions and iterations during the creation of a video game, does not imply the relevance and the meaningfulness of those contingencies for the activity and the collective product. If we fail to question this opportunity collectively, if not democratically, by considering the points of view of the different actors, we risk undermining their commitment. Indeed, it is not well known that digital work always calls for more organizational work (Bidet *et al.* 2017). By multiplying opportunities to revise and experiment, by allowing flexible distribution of activity within a network of human and non-human actors, in space and time, the rise of digital technologies also requires more work to weave the continuity of activity, both individual and collective. Multiplying situations of multi-activity, arbitration and interface increases the organizational work required at all hierarchical levels to both organize and make activities mutually intelligible.

12.6. References

Alter, N. (2010). *Donner et prendre. La coopération en entreprise.* La Découverte, Paris.

Auray, N. (2003). Le sens du juste dans un noyau d'experts. Debian et le puritanisme civique. In *Internet, une nouvelle utopie limitée. Nouvelles régulations, nouvelles solidarités*, Conein, B., Massit Lolleat F., and Proulx S. (eds). Presses de l'Université de Laval, Quebec.

Bernoux, P. (2011). Reconnaissance et appropriation : pour une anthropologie du travail. *Esprit*, 10, 158–168.

Bessy, C. and Chataurounaud, F. (1995). *Experts et faussaires. Pour une sociologie de la perception.* Métaillé, Paris.

Bidet, A. (2008). L'homme et l'automate. L'écologie élargie du travail contemporain. *Sociologie du travail*, 50(3), 372–395.

Bidet, A. (2011a). *L'engagement dans le travail. Qu'est-ce que le vrai boulot ?* Presses Universitaires de France, Paris.

Bidet, A. (2011b). La multi-activité, ou le travail est-il encore une expérience ? *Communications*, 89(2), 9–26.

Bidet, A. (2017a). Être "contraint de s'organiser" : la multi-activité, entre situations de travail et formes sociales. In *Quand travailler c'est s'organiser. La multi-activité à l'ère du numérique*, Bidet, A., Datchary C., and Gaglio G. (eds). Presses des Mines, Paris.

Bidet, A. (2017b). Actualiser les images du travail. In *Penser le travail pour penser l'entreprise*, Favereau O., Bidet A., Le Gall J.-M., Lopes H., Roger B., and Seignour A. (eds). Presses des Mines, Paris.

Bidet, A. and Boutet, M. (2013). Pluralité des engagements et travail sur soi. Le cas de salariés ayant une pratique ludique ou bénévole. *Réseaux*, 6(182), 119–152.

Bidet, A. and Porta, J. (2016). Le travail à l'épreuve du numérique. *Revue de droit du travail*, 5, 328–334.

Bidet, A. and Vatin, F. (2016). Travailler, c'est produire : activité, valeur et ordre social. In *L'activité en théorie – regards croisés sur le travail*, Dujarier, M.-A., Gaudart, C., Gillet A., and Lenel P. (eds). Octarès Editions, Toulouse.

Bidet, A., Datchary, C., and Gaglio, G. (eds) (2017). *Quand travailler c'est s'organiser. Le travail à l'ère numérique*. Presses des Mines, Paris.

Boltanski, L. and Chiapello, E. (1999). *Le nouvel esprit du capitalisme*. Gallimard, Paris.

Chateauraynaud, F. (1997). Vigilance et transformation. Présence corporelle et responsabilité dans la conduite des dispositifs techniques. *Réseaux*, 5(85), 101–127.

Clot, Y. (2010). *Le travail à coeur. Pour en finir avec les risques psychosociaux*. La Découverte, Paris.

Dewey, J. (1987). Democracy in education. *The Later Works of John Dewey, 1925–1953*. Southern Illinois University, Carbondale.

Dodier, N. (1995). *Les hommes et les machines. La conscience collective dans les sociétés technicisées*. Métailié, Paris.

Hughes, E.C. (1996). Le travail et le soi. *Le regard sociologique*. Editions de l'EHESS, Paris.

Lhuilier, D. (2016). L'activité, dans et au-delà du monde du travail. In *L'activité en théorie – regards croisés sur le travail*, Dujarier, M.-A., Gaudart, C., Gillet A., and Lenel P. (eds). Octarès Editions, Toulouse.

Naville, P. (1963). *Vers l'automatisme social. Problèmes du travail et de l'automation*. Gallimard, Paris.

Rot, G. and Vatin, F. (2017). *Au fil du flux. Le travail de surveillance-contrôle dans l'industrie chimique et nucléaire*. Presses des Mines, Paris.

Schwartz, Y. (2000). *Le paradigme ergologique ou un métier de philosophe*. Octarès, Toulouse.

Vatin, F. (1987). *La fluidité industrielle. Essai sur la théorie de la production et le devenir du travail*. Méridiens-Klincksieck, Paris.

Zimmermann, B. (2017). Entre valorisation de soi et mise à l'épreuve de soi : les dynamiques paradoxales de l'autonomie. *Formation emploi : revue française de sciences sociales*, 139(3), 91–104.

13

Is the Obsolescence of the Skills of Older Employees an Inevitable Consequence of Digitalization?

Digital technology characterizes new forms of work organization and requires employees to develop new skills. However, this development is not self-evident, especially for certain "vulnerable" populations, such as older employees. In this chapter, after a presentation of the implications of aging in the context of an activity mediated by a technical tool, we will present the results of research conducted among SNCF controllers. By looking at their activity "in the making", we have highlighted the conditions promoting the development of skills suitable for the use of a technical tool.

13.1. Introduction

The large-scale integration of digital technology in various forms and in most professional spheres has paved the way for profound transformations and speculation about the future of work (Gomez 2018). The literature, which is abundant on this subject, highlights the extent to which the use of digital technology can have an effect on employment, improving certain aspects (autonomy, versatility) and weakening others (control, deskilling). One of these is of particular interest to us and will be the subject of this chapter: the links between technologies and skills (Bailey *et al.* 2010). Studies conducted using the "skill-biased technological change" approach show a shift in job creation from the most skilled to the least skilled (Card and DiNardo 2002), thus revealing a "digital divide" (Gualtieri 2015). Other

Chapter written by Florence CROS, Marc-Eric BOBILLIER CHAUMON and Bruno CUVILLIER.

researchers (Van Loo 2007; Rohwedder and Willis 2010) show that these transformations carry a risk of the technical and economic obsolescence of skills, which can be heightened among older employees (Thijssen and Walter 2006).

In the current societal context, the issue of skills obsolescence has become central. Indeed, as demographic trends are moving towards an aging population, employment policies must be re-discussed in order to take care – socially and economically – of this part of the population. The extension of working life up to 64 years of age has therefore been advocated. However, the changes are leading to profound transformations that make older employees more vulnerable. It is therefore essential to think jointly about the issue of aging and the development of technologies, in order to enable older workers to remain active and healthy at work.

In this chapter, and based on a study conducted among SNCF controllers (a French railway transport company), we will question the relationship between aging and technology in the workplace by examining the conditions for developing the appropriate skills for using a technical tool, Accelio[1].

13.2. Aging, work, technologies and skills obsolescence: theoretical elements

The study of the relationship between aging and technology is, as Teiger (1989) described it, "an old problem in a recent context". In this sense, much research has already been conducted to characterize these relationships. Here, we present the main results that point to aspects to which we should pay particular attention in our study.

13.2.1. *The effects of aging*

Commonly, aging is associated with the idea of sudden and inevitable decline. This perspective is that of a biological approach based on cellular degeneration and resulting in a weakening of human functioning and cognitive performance (Salthouse 2003). This cognitive decline has a direct effect on information processing and the ability to inhibit information (Hasher and Zacks 1988). The developmental approach to aging (Goulet and Baltes 1987) proposes that we consider advancing in age as an alternation of gains and losses, by means of the experience available to each individual. According to Delgoulet *et al.* (2014), aging is "a dual movement of declines and gains, the direction, magnitude, and occurrence of which vary from one person to another" (2014, p. 690). This variability can be

1 Accelio is a terminal that allows the controller to be synchronized with the SNCF's central systems and thus allows access to traffic information (timetables, fares, etc.).

explained by the importance given to cultural components that allow individuals to limit or even counter the effects of age. The constant interaction between these two components thus defines aging as a nonlinear and developmental process.

13.2.2. *Aging associated with work*

Studies in the humanities and social sciences over the past 20 years have contributed to the enrichment of a three-pronged model of aging associated with work (Teiger 1989): aging "through" work, aging "in relation to" work and aging "in" work. These three approaches have in common the link between aging and work organization (David *et al.* 2001). Indeed, the work situations experienced by aging employees have an effect on the evolutions and involutions of the individual; they can induce or even put an emphasis on declines, have a protective effect on health (Gaudart and Weill-Fassina 1999) or even reveal declines (Delgoulet and Volkoff 2014). Finally, studying the effects of aging in the workplace means taking into account not only the decline in functional abilities but also their maturation and development through lived experience. Consequently, the effects of aging on work activity are multiple and complex to fully understand.

13.2.3. *From the effects of technology on activity to the issue of the obsolescence of older workers' skills*

The deployment of technical solutions in the professional sphere is a promise of performance, modernity and progress. Beyond these economic aspects, these technical solutions, which can be envisaged in various forms, also represent a vector for improving the quality of life at work by developing human abilities (Oosterlaken 2009). At the same time, some authors point out the harmful effects of these technologies, which can be described as disruptive (Bobillier Chaumon 2016), as they require profound changes in work methods: loss of autonomy, flexibility, new practices, new skills, etc. (Greenan 2003; Aubert *et al.* 2006). As adaptability is becoming a key skill, some employees may find themselves in difficulty. This is particularly the case for older workers who lose this ability as they get older (Teiger 1995). Technological change can therefore expose older employees to a major negative shock (Friedberg 2003; Daveri and Maliranta 2006). The qualifications that previously protected these employees may no longer suffice (Czaja and Sharit 1993), at the expense of their experience, which is a guarantee of expertise in the profession. The risk of skills obsolescence seems to be a major threat among older employees. To remain employable, employees must constantly update their knowledge and skills. The notion of skills obsolescence, as defined by Rosen (1975), refers to two aspects: negative changes over time and the provision of knowledge by the social environment. Consequently, obsolescence results, on the one hand, from advancing age and, on the other hand, from the technologies deployed in the activity. We therefore speak of technical and economic obsolescence, both of which are assessed by various methods (Van Loo and de Grip 2001; de Grip and Van Loo 2002) that consist of measuring productivity or the probability of losing one's job. Although these methods have been found to be robust, they do not allow us to understand the dynamics between the activity of older workers and the use of technology.

13.3. Question and methodology

As these various theoretical elements show us, aging is a complex phenomenon, composed of cultural and biological aspects. When it is studied in cases of technology use, then other difficulties arise, as evidenced by the increased risk of skills obsolescence among older employees highlighted in the literature. However, developmental approaches lead us to question this risk of skills obsolescence among older workers in a different way. Indeed, this notion, mainly studied in the field of economics, relies on quantitative indicators (such as productivity) at the expense of those that can be highlighted by the observation and analysis of the activity carried out by these older employees (Clot 1999). However, aging is not only a biological process of involutions; it is to be understood in terms of development and is therefore also made up of learning, particularly through the environment, the collective, experience, etc. (Clot 1999).

It is therefore from a psychosocial perspective that we wish to question the notion of skills obsolescence. In order to do so, we will examine the socio-technical conditions that allow older employees to continue their activity when it is mediatized by a technological tool. To answer this question, we looked at the activity of SNCF controllers.

13.3.1. *Study background and methods*

Our intervention took place in an establishment with a team of 105 controllers (subsequently named ASCTs[2]), supervised by a unit manager (DUO), an assistant manager (RDUO) and five local managers. In a context marked by profound organizational and technological transformations, the objective of the intervention was to understand the working methods of young and older employees in order to understand which job skills were transmitted and how, if at all.

2 ASCT: *Agent de Service Commercial Train* (a commercial train service agent).

The ASCT's job is done alone, on board trains. ASCTs are in charge of the safety of trains and passengers, as well as of their comfort and the control of their tickets. Thus, each user must be able to travel regularly, in optimal safety conditions. Within the framework of this study, we met five controllers, three local managers and the RDUO. The average age of the ASCTs was 47; they had been employed by the company for an average of 28 years.

In order to capture the "older employees'" relationships with technology during the course of the activity, we deployed a methodological approach using triangulation. Initially, this involved conducting interviews with each professional, with the aim of understanding the – prescribed – job of controller, from different points of view. The second step was to observe the activity from a local point of view by following the five controllers on board the trains on a daily basis in order to understand the activities carried out. Finally, the last step was the explanation, based on previously observed sequences. The objective of this final phase of the study was to access the experience (Vermersch 1994) of employees during work situations that we had previously identified. All the data collected was then analyzed, first individually and then on an interdisciplinary level.

13.3.2. *Engeström's activity system model (1987)*

The activity system model, developed by Engeström (1987) on the basis of work carried out by psychologists of the Russian historical–cultural current (Leontiev, Vygotsky), has the major advantage of conceiving of activity in a systemic and global perspective. In this sense, we will use it both as a model to understand the activity and then to analyze it. According to Engeström (1987), the six elements that constitute this model (subject, instruments, object, rules, community, division of labor) are interdependent (Figure 13.1).

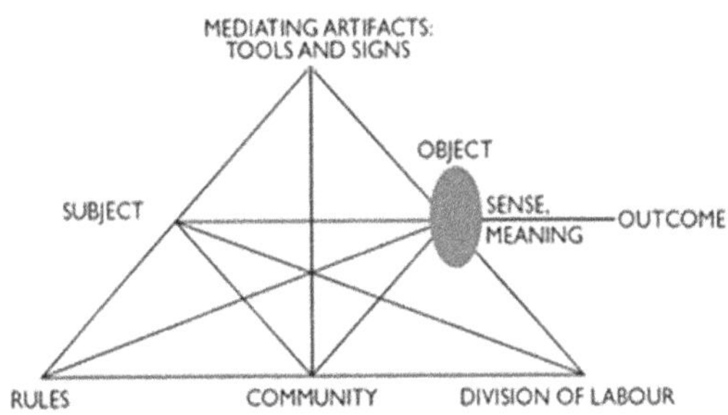

Figure 13.1. *Activity system model (Engeström 1987, p. 78)*

Thus defined, the activity system makes it possible to take into account the context within which the professional practices of individuals emerge, develop, are inscribed or even contradict each other when two extremes are in tension. Kuutti (1996, p. 34) considers that tensions are manifestations of "problems, ruptures, breakdowns" that interrupt the normal flow of work. However, they are essential because they are potential sources of development if the individual manages to overcome them. The interest of this model within the framework of our study is therefore to be able to simultaneously consider the variations in the experience, qualifications and skills of individuals and their differentiated vision of the object of the activity.

13.4. Main results

The analysis of the data collected enabled us to highlight several results that provided answers to our question, namely whether older employees manage, in the context of an activity mediated by technology, to develop the appropriate skills, and under what conditions? We will first show that the ASCT profession is not defined according to the same age-related skills and quality criteria, leading to tensions within the ASCT activity. In a second step, we will see how older ASCTs managed to overcome these tensions in order to continue their activity.

13.4.1. *The ASCT profession: a perception through the prism of age*

The ASCT profession is defined by four main principles: the "4 S's" which are *sécurité* (security), *sûreté* (safety), *service* (service) and *sauvegarde des recettes* (revenue protection). These four principles are variously distributed and invested in during the ASCT's activity. They are therefore linked to the way ASCTs perceive their job and what makes it a quality job.

13.4.1.1. *Skills, recognition and quality of work*

Our analyses have shown us a divergence in the perception of the profession and the way it is perceived, with regard to age. For example, the youngest ASCTs considered safety and security to be the most important aspects of their job, although the actual workload related to these aspects was smaller. The older ASCTs described the core tasks being those related to safeguarding revenue (ticket control, regularization, speaking with customers). This divergence in the perception of the profession can be analyzed in terms of divergent quality criteria at work (Clot 2008). Indeed, for young ASCTs, safety and security requires specific skills, which, when implemented, were visible to passengers. They are a vector of recognition. If ASCTs carried out safety- and security-related tasks, then they felt that they had done their job well. Conversely, older ASCTs perceived more recognition from travelers when they carried out control, regularization and information tasks through the relationships they created, "based on trust on both sides" (Romain 56). It is in this service relationship that quality work for older ASCTs could be found.

The differentiated appreciation of the controller's job was reinforced by the qualities of the ASCT's technical work tool, Accelio (see Figure 13.2).

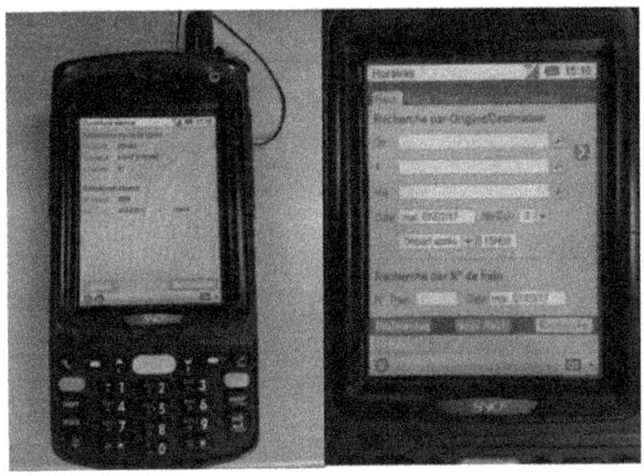

Figure 13.2. *Accelio. For a color version of this figure, see www.iste.co.uk/bobillierchaumon/digital.zip*

The ASCTs agreed on the slowness of this tool, the difficulty it caused them and the consequences of it, such as the "frustration" and "shame" it engendered. However, older ASCTs noted that, despite a lack of training on the tool, it had contributed to improving working conditions by reducing the cognitive load associated with their activity by:

– limiting the risk of error when drawing up a report; consequently, a discussion with customers is more frequent;

– allowing each officer to inform on board which train he or she is on; this contributes to the reduction of apprehension felt when an ASCT is alone on board the train, promoting contact with the other professions in charge of moving the train;

– allowing room for maneuver in relation to the role's restrictions and thus demonstrating flexibility towards passengers, calming people, defusing a possible conflict, etc.

These different aspects related to the use of Accelio point to the role of interpersonal relations in the ASCT profession. They are diverse: between colleagues, with hierarchy and travelers. What can be retained is that the "controller" collective has largely weakened, with the ASCTs speaking of "dislocation of the collective" following the great strike of 1995 in France, in favor of an inter-profession collective created elsewhere and otherwise, and more particularly invested in by the older ASCTs. This new form of collective made up of controllers, security support agents, stopover agents, members of the operational center, etc. was perceived by the older ASCTs as "teamwork" (Romain 56 years old), working towards a common goal and therefore supporting the activity of the controllers.

13.4.1.2. *Points of divergence in an ASCT's activity*

These results, based on the analysis of interviews conducted with ASCTs and various managers, allowed us to understand the main characteristics of the profession and to establish three observations:

– younger ASCTs and older people did not perceive their profession in the same way: there was a divergence of purpose;

– the expectations of the profession for young ASCTs were not the same as for older ones;

– inter-profession relationships were central for older ASCTs who saw themselves as contributing to teamwork.

We have seen that the ASCTs did not explicitly express the role of Accelio in their daily activity, even though it was their main working tool. However, after vertical and then transversal analysis, we noted that it was the mediating element of the main tensions identified within the ASCTs' activity. Indeed, since it was not as effective as desired, it sometimes put the ASCTs in difficulty. While the younger ASCTs disinvested in its use in favor of other more rewarding tasks that did not require the use of Accelio, the older ASCTs maintained this use and managed to carry out quality work.

13.4.2. *Overcoming tensions to develop one's activity*

The main contradiction that we were able to identify from the analysis of ASCT discourses is the one that emerged between the subject–instrument–object extremes of the activity system, modeled by Engeström. It can be translated in these terms: Accelio did not allow ASCTs to do quality work for tasks related to income protection. Indeed, many malfunctions were detected and the activity of the ASCTs was constrained by Accelio, which occasionally prevented them from doing their work "well".

During activity monitoring, we observed these "bugs" on several occasions (start-up, initialization, collection, etc.) which caused the worker to operate in a degraded mode. It was these situations that we were particularly interested in during the explanatory interviews, in order to understand the subjective processes underlying the overcoming of this tension. Two different strategies implemented by older ASCTs were identified:

– In the first case, Julien (57) was unable to bill a ticket; his debit card reader did not work. He explored his field of possibilities by changing his strategy (several collection attempts, help from a colleague, another reader, another debit card). In spite of this, he was unable to take the payment. He then changed the initial object of his activity, a regularization activity, to one involving discussion. Accelio's malfunctions thwarted Julien's activity and led us to look at the motives for the activity, that is, Julien's motivations for redirecting his activity. The first of them related to the reliability of the tool, linked to his experience with it. As dysfunctions were recurrent, he deployed room for maneuver in order to regain control of the situation and allow the customer to travel with confidence. The second motive related to the quality of the work, which Julien defined as the possibility of deploying "all the means" to achieve the result, that is, "control" without being "scorned when I want to do a control operation and do it thoroughly".

– In the second case, Romain (56 years old) was confronted with a traveler who did not have a ticket. This situation caught our attention during the activity follow-up because instead of using Accelio, Romain used the ticket book. The latter can be used in two situations: when Accelio has broken down and when the passenger makes a false declaration of identity (legal requirement). Romain did not find himself in either of these situations; he explained that he did it "out of habit" and "because it is faster than with Accelio". In the course of the explanation, we came back to this point and we realized that the misuse of Accelio's use, a strategy adopted by Romain, was underpinned by a motive other than habit. In fact, the officer explained that the objective "is to resolve the situation as quickly as possible" in order to stay with the offender as little time as possible because "it is extremely dangerous to stay in front of someone who is angry". This misuse of Accelio revealed both the status that the tool had for Romain and the level of appropriation that it was the object of. It was while trying to integrate Accelio into his professional practices that Romain realized that it did not allow him to comply with the regulatory injunctions that recommended "staying for as short a time as possible".

13.5. Discussion and conclusion

In this chapter, we proposed to see whether or not the obsolescence of skills is an inevitable consequence of digitalization when it is carried out during the activity of older employees. This study carried out among SNCF controllers showed that older employees were able to develop suitable skills for using Accelio, under certain conditions.

13.5.1. *Accelio, a vector of recognition for the ASCT profession?*

We have shown, through two use cases, that older ASCTs managed to use Accelio, their main working tool, which replaced the ticket book. This tool reported numerous malfunctions, which sometimes hindered or even prevented the "normal" activity of the controllers. Variable uses were then observed among the ASCTs; the youngest ones abandoned this tool, while the older ones continued to use it. Accelio therefore did not have the same status in the activity of young ASCTs and older ones. Bobillier Chaumon *et al.* (2006) explain, based on the results of a study conducted in the banking sector, that the tool's role and contribution condition its uses. Thus, the more the use of the tool is justified, makes it efficient and meets expectations, the more it is acceptable and accepted by the subjects. According to Bobillier Chaumon *et al.* (*ibid.*, p. 11), "acceptance [of the system] is therefore strongly linked to the representations [that each agent] makes of their own skills (technical, commercial, etc.) and professional objectives and the types of relationship with the system that this presupposes". The way in which the tool was used therefore depended on the image that the ASCT wished to give, namely that of not wanting to put themselves in a weak position in order to maintain a positive identity with travelers. And it is because they managed to recognize themselves in their vision of the profession that older ASCTs found the means to act effectively, by renewing the goals of their actions.

13.5.2. *Diverting to work better*

The act of diverting is central to the activity of ASCTs: diverting the object and diverting the use. Prescription is therefore, as Clot (1999) has shown, necessary but not sufficient. It is meant to be questioned and improved upon. These deviations from the prescription (Bourrier 1999) are made in relation to the existing situation and through adaptation. They are also the manifestation of the autonomy of ASCTs who were able to deal with the unexpected by jointly mobilizing prescription and their ingenuity. In this sense, these deviations – indicators of the development of employees' activity – took the form of what Ombredane and Faverge (1955) called a catachresis. According to Clot and Fernandez (2012, p. 327), catachresis is a general characteristic of the actual psychological development of the individual. When developing a catachresis, the subject puts the tool at the service of their present activity in order to carry out the unobservable activity. The tool is then not in itself an instrument of action; it becomes one in a situation that awakens them (Rabardel 1995). From this perspective, Accelio enabled older ASCTs to "do" their work well despite the characteristics of the situations, their professional logic and the quality criteria that guided them.

To be able to carry out these diversions, the agents integrated all the characteristics of Accelio as well as the effects of its use according to the situation. Thus, Romain noted that the use of Accelio in "risky" verbal situations endangered the ASCTs. In order to protect himself, Romain diverted the use of Accelio in favor of the ticket book, which he used to distance himself from the offender.

Finally, without reappropriation of the tool through misappropriation, circumvention and non-use, there is no appropriation. However, one only appropriates what one can finally recognize oneself in (Bianchi and Kouloumdjian 2006). To divert is to put the "self" into the tool so that it meets our expectations, our requirements, the functionalities that we would like to attribute to it and even so that affinities are created with respect to what it allows.

13.6. References

Aubert, P., Caroli, E., and Roger, M. (2006). New technologies, organization and age: Firm-level evidence. *The Economic Journal*, 116, 73–93.

Bailey, D.E., Leonardi, P.E., and Chong, J. (2010). Minding the gaps: Understanding technology interdependence and coordination in knowledge work. *Organization Science*, 21(3), 713–730.

Bianchi, J. and Kouloumdjian, M.-F. (1986). Le concept d'appropriation. In *L'espace social de la communication*, Laulian, A.-M., Bianchi, J., and Kouloumdjian, M.-F. (eds). Retz/CNRS, Paris.

Bobillier Chaumon, M.E. (2016). Acceptation située des TIC dans et par l'activité : premiers étayages pour une clinique de l'usage. *Psychologie du travail et des organisations*, 22(1), 4–21.

Bobillier Chaumon, M.E., Dubois, M., and Retour, D. (2006). L'acceptation du changement technique : le cas des nouvelles technologies dans le milieu bancaire. *Psychologie du travail et des organisations*, 12(4), 247–262.

Bourrier, M. (1999). *Le nucléaire à l'épreuve de l'organisation*. PUF, Paris.

Card, D. and DiNardo, J.E. (2002). Skill-biased technological change and rising wage inequality: Some problems and puzzles. *Journal of Labor Economics*, 20, 733–783.

Clot, Y. (1999). *La fonction psychologique du travail*. PUF, Paris.

Clot, Y. (2008). *Travail et pouvoir d'agir*. PUF, Paris.

Clot, Y. and Fernandez, G. (2012). Mobilisation psychologique et développement du "métier". In *Traité de psychologie du travail et des organisations*, Bernaud, J.-L. and Lemoine, C. (eds). Dunod, Paris.

Czaja, S. and Sharit, J. (1993). Age differences in the performance of computer-based knowledge, personality and motives. In *International Review of Industrial and Organizational Psychology*, Cooper, C.L. and Robertson, I.T. (eds). Wiley, Chichester.

Daveri, F. and Maliranta, M. (2006). Age, technology and labour costs. Discussion Paper, 1010, The Research Institute of the Finnish Economy, Helsinki.

David, H., Volkoff, S., Cloutier, E., and Derriennic, F. (2001). Vieillissement, organisation du travail et santé. *Pistes*, 3(1).

Delgoulet, C. and Volkoff, S. (2014). Vieillissement, pénibilité et travail. *Pathologie professionnelle et de l'environnement*, 9(2), 1–8.

Delgoulet, C., Volkoff, S., Caron, L., Caser, F., Jolivet, A., and Théry, L. (2014). Conditions de travail et maintien en emploi des seniors : enjeux d'un "décloisonnement" des approches et des pratiques : quelques enseignements issus de monographies d'entreprises françaises. *Relations industrielles*, 69(4), 687–708.

Engeström, Y. (1987). *Learning by Expanding: An Activity-theoritical Approach to Developmental Research*. Orienta-Konsultit, Helsinki.

Friedberg, L. (2003). The impact of technological change on older workers: Evidence from data on computer use. *Industrial and Labor Relations Review*, 56, 89–120.

Gaudart, C. and Weill-Fassina, A. (1999). L'évolution des compétences au cours de la vie professionnelle : une approche ergonomique. *Formation emploi*, 67(1), 47–62.

Gomez, P.Y. (2018). Les transformations du travail dans l'économie numérique. *Entreprendre and innover*, 2(37), 8–17.

Goulet, L.R. and Baltes, P.B. (1970). *Life-span Developmental Psychology: Research and Theory*. Academic Press, New York, USA.

Greenan, N. (2003). Organizational change, technology, employment and skills: An empirical study of French manufacturing. *Cambridge Journal of Economics*, 27, 287–316.

de Grip, A. and Van Loo, J. (2002). The economics of skills obsolescence: A review. In *The Economics of Skills Obsolescence*, de Grip, A., Van Loo, J., and Mayhew, K. (eds). JAI Press, Bingley.

Hasher, L. and Zacks, R.T. (1988). Working memory, comprehension and aging: A review and a new view. In *The Psychology of Learning and Motivation*, Bower, G.H. (ed.). Academic Press, New York, USA.

Kuutti, K. (1996). Activity theory as a potential framework for human-computer interaction research. In *Context and Consciousness: Activity Theory and Human Computer Interaction*, Nardi, B.A. (ed.). The MIT Press, Cambridge, USA.

Ombredane, A. and Faverge, J.M. (1955). *L'analyse du travail. facteur d'économie humaine et de productivité*. PUF, Paris.

Oosterlaken, I. (2009). Design for development: A capability approach. *Design Issues*, 25, 91–102.

Rabardel, P. (1995). *Les hommes et les technologies. Approche cognitive des instruments contemporains*. Armand Colin, Paris.

Rohwedder, S. and Willis, R.J. (2010). Mental retirement. *Journal of Economic Perspectives*, 24(1), 119–138.

Rosen, S. (1975). Measuring the obsolescence of knowledge. In *Education, Income and Human Behavior*, Thomas Juster, F. (ed.). McGraw Hill, New York, USA.

Salthouse, T.A. (2003). What and when of cognitive aging. *Current Directions in Psychological Science*, 13, 140–144.

Teiger, C. (1989). Le vieillissement différentiel dans et par le travail : un vieux problème dans un contexte récent. *Le travail humain*, 52(1), 21–56.

Teiger, C. (1995). Penser les relations âge/travail au cours du temps. In *Le travail au fil de l'âge*, Marquié, J.C., Paumès, D., and Volkoff, S. (eds). Octarès, Toulouse.

Thijssen, J. and Walter, E. (2006). Identifying obsolescence and related factors among elderly employees. Conference paper [Online]. Available at: https://www.ufhrd.co.uk/wordpress/wp-content/uploads/2008/06/50-2_thijssen_walter.pdf.

Van Loo, J.B. (2007). The speed of obsolescence: Evidence from the Dutch public sector. Paper, International Research Conference in the Americas of the Academy of Human Resource Development, Indianapolis [Online]. Available at: https://files.eric.ed.gov/fulltext/ED504838.pdf.

Van Loo, J., de Grip, A., and De Steur, M. (2001). Skills obsolescence, causes and cures. *International Journal of Manpower*, 22, 121–137.

Vermersch, P. (1994). *L'entretien d'explicitation*. ESF, Paris.

14

Are Work Collectives and Digital Exposure Compatible?

For almost half a century, the work sector has been digitalizing: technologies (computers, smartphones, software, etc.) are omnipresent in many companies. Engineering, a context of hard sciences and mathematics, is of course no exception to this phenomenon. However, technology is based on a binary and formalized language that does not take into account all the needs and capacities for expression and criticism that are specifically human. Here, we seek to recognize and characterize, more precisely, the role and consequences of the use of so-called collaborative digital tools on collective activity.

14.1. Collective activity: major developments

Collective activity needs to be preserved and developed at work because it is a source of organizational performance, quality of work, as well as health (Caroly 2010; Clot 2010): it carries the possibilities of co-construction and mutual interdependence of employees, as well as their power to act, in its individual and collective form (Clot 2008). Collective activity corresponds to the setting in motion of collective work and of the work collective(s) that carry it out (Caroly 2012).

Collective work (Caroly and Barcellini 2013) corresponds to explicit coordination activities and adjustments of tasks and missions according to the risks of the activity. It refers to the articulation of competences according to common operative frames of reference (de Terssac and Chabaud 1990; Leplat 1991, 2001), that is, functional, shared and common representations of the activity in progress.

Chapter written by Pauline CROUZAT and Marc-Eric BOBILLIER CHAUMON.

Different forms of collective work were defined by de la Garza and Weil Fassina (2000): joint action, joint activity, collaboration, cooperation, help and mutual assistance. The latter depend not only on the needs of the actors throughout their tasks but also on the nature and organization of the projects. The main socio-cognitive resource is found in the opportunities for synchronization, whether operative or cognitive (Grosjean and Lacoste 1999). Operational synchronization corresponds to operational articulation, that is, "work through which [...] information is constructed, actions are coordinated, and activity is organized and reorganized according to the constant risks of work" (Grosjean and Lacoste 1999, p. 168). Cognitive synchronization corresponds to the articulation of trajectories and focuses on "the target of the object" (Grosjean and Lacoste 1999, p. 170). It provides a clear picture of the project's current situation in order to determine the objectives to be achieved, the means to achieve them, and to review the strategy in light of changes, constraints and even uncertainties.

The work collective (Clot 2008; Caroly 2010, 2011) is made up of a group of actors with a tacit mutual understanding of the activity, who share implicit rules that guide their work (job- or function-type reference systems). This work collective (collective body) is structured and consolidated through confrontation, dialogue (Litim and Kostulski 2006), controversy and debate. Above all, it is structured over time: it writes the history of the group and even of the professions that constitute it. It is also a commonly accepted reference that enables the evaluation and qualitative judgment of the work carried out (Dejours 1993), based on collectively accepted references and benchmarks.

Within certain matrix organizations, particularly in industrial engineering, collectives, which are now cross-functional (Poret 2015), tend to become more fragile given the organization of work that has evolved, particularly towards a project mode: often, work is no longer organized around professions but rather around ongoing projects. Mutual identification, still based on professions from a few years ago, is today essentially based on functions, tasks or roles (often changing). This evolution tends to complicate the spontaneous and tacit understanding that characterizes a work collective in the traditional sense (Clot 2010). The professional genre (Clot 2000), that is, the set of evaluations shared among professionals, tacitly conditions and regulates personal activity (Clot 2000). It is a body of rules, practices, norms and even implicit strategies shared by peers. In the context studied, these genres are struggling to structure themselves around sustainable reference frameworks (as are the foundations of job-type reference frameworks). In the same way, all reference frameworks, whether operational or functional, are essentially visual and are structured and formalized on the basis of, and around, the technologies available to employees.

14.2. Engineering: a highly digitized environment

Engineering is a highly digitized sector where a large amount of information circulates. There are many initiatives and digital tools to improve communication, facilitate exchanges and thus improve cooperation and collaboration: e-mails, video conferencing, digital touch screens (visual management), clouds, 3D model projectors, report software, presentation software , as well as all shared files (forms, tables, text files, etc.). These tools all have one thing in common: the visual is now a preferred medium for presenting and sharing information.

They are not only used by professionals remotely but have also become a reference in terms of information support (meetings, presentations, appointments, mutual monitoring of work status) and monitoring of activity, both collective and individual. The deciphering of the content of these tools is essentially "fast", as is the circulation of information. Beyond e-mail, which is already old, the formats of the tools tend to favor indicators, pictograms, diagrams, images and sounds to the detriment of argued and written language (which requires interpretation and the time to do to it). This raises the question of collaboration, which is based mainly on these uses and on the dialogue and oral dimension that surrounds them. Do these tools allow for dialogue and debate around the activity being carried out, as well as around mutual practices? What is the impact of these tools in terms of collective activity? More specifically, what is their role in terms of collective work, and, above all, of the work collectives that implement them?

14.3. Problem of the study

Collaboration is highly instrumented and favored in view of the high number of schemes with a cooperative objective. However, some industrial engineering employees we met told us that they did not really know which team they belonged to. At the crossroads of project teams, administrative teams and their functions, it seems that the cooperative reality does not always correspond to the formal and publicly disseminated functional organization chart. Similarly, according to the actors, work tends to become progressively more individualized as collaborative technologies multiply (Crouzat 2019).

Our objective was therefore to verify how the latter instrumentalized collective activity. More specifically, we sought to find out whether the digital collaborative tools, made available to employees, impacted their collective activity and, in particular, whether and how their use impacted collective work and/or the work collective. It was by relying on the clinical investigation of use (Bobillier Chaumon and Clot 2016), that is, by questioning and characterizing the actual use of the tools available, that we carried out our analysis.

14.4. The methodology used

Here, we were not interested in tools certain to specific activities (design, industrial drawing, accounting, etc.). We focused on so-called *collaborative* management and communication tools in digital form. Chiapello and Gilbert (2012) define a management tool as "a delimited set of organizational objects". A management tool always has an organizational purpose (Chiapello and Gilbert 2012). It exists only when incorporated into a work system or device. It is not just an indicator. It must be: associated with a managerial logic; associated with elements of the organization (procedures, meetings, written reports); related to others; used by several actors; and have the ability to facilitate communication between people. For Chiapello and Gilbert (2012), the sphere of these tools is "as much social as technical". These tools were used by all engineering professionals and all had one thing in common: they were deployed to improve communication and facilitate cooperation between employees. Their prescribed use was therefore to improve collaboration.

Here, we only focus on digital management and communication tools (our thesis research also included many other non-digital devices). We chose to observe seven tools (see Table 14.1), which were the most used.

	Apparent and known functional properties	**Prescribed uses**
Electronic messaging	Communication	Promote remote communication
Diary manager	Coordination/Collaboration	Promote coordination of meetings
Video conferencing software	Coordination/Communication	Enable remote meetings, promote remote communication
Instant messaging	Communication	Promote remote communication
Corporate social network	Information/Collaboration	Foster links and information sharing among employees
Shared folders	Centralized access to information	Centralize a team's information in a single location for easy access
PowerPoint	Structuring and presentation of information/Communication	Facilitate presentations through visual references

Table 14.1. *List and characteristics of digital tools observed (Crouzat 2019)*

E-mail corresponds to the e-mail management software used in the company. The diary manager was a tool that enabled visibility of the availability of others, and sharing of one's own, for making appointments and organizing meetings. The video conferencing software allowed organization of appointments and meetings online. It was thus possible to speak, as well as to see (via an integrated camera), the interlocutors remotely. This software also allowed screen sharing and presentation of documents in real time. The company's social network was a digital "networking" tool intended to facilitate a shared and collaborative operation within the company. It offered internal company employees the opportunity to post articles or notes, comment on them, and create individual pages or blogs, communities and forums that could be accessed. Shared folders were servers that allowed storage and collective access to documents that had been completed or were in the process of being completed. In particular, they allowed several contributors to have access to a shared document in order to modify or consult it. The Microsoft PowerPoint software was a tool used almost daily by actors: it was the shared presentation format (medium for presentations, meetings and discussions, meeting minutes). Its nature, as a slideshow, promoted, in particular, the common visualization of the information it contained (by projecting it on a large screen, which many rooms were equipped with).

At first glance, the apparent functional properties of these tools and the associated prescribed uses (Table 14.1) appeared to facilitate communication and collaboration. In order to analyze the concrete consequences of these tools on collective activity, we therefore focused on observations made in situations of actual use, with a grid that we developed from the psychological and socio-cognitive resources of collective activity, known in the scientific literature. This observation grid (Table 14.2) allowed us to take into account verbal elements and behaviors corresponding to the manifestations listed in the specific observation axes. The final grid (presented without details of the specific manifestations observed) is as follows.

The grid is based, primarily, on the work of Wisner (1993) and Caroly and Barcellini (2013) who defined the psychological and socio-cognitive resources of collective activity. Our observations lasted a total of 94 hours and involved a total of 1,020 engineering actors from several departments (quality, avionics design, research and development, etc.), aged between 21 and 65 years, all statuses and genders combined. These observations concerned both individual use (using so-called collaborative tools) and collective use (visual screens).

COLLECTIVE ACTIVITY COMPONENTS AND RESOURCES	COLLECTIVE WORK				WORK COLLECTIVE			
	Operative synchronization	Cognitive synchronization			Conflict and regulation management in action	Development and/or re-development of work and of the profession	Debate on the quality of work	Contribution to the common history of the activity
		Concerning the current situation	Concerning understanding and mutual visibility	Through the sharing of knowledge				
INDICATORS: OBSERVABLE MANIFESTATIONS (usage content: verbal expressions, types of exchanges and behaviors)	Adjustments, distribution, coordination and planning of tasks	Chronological progress of the tasks and their realization (realized tasks, future tasks, delays, difficulties encountered). Clarification of tasks and objectives to be achieved	Collective dissemination of commented visual information for clarification and explicitation of the work. Making visible and/or presenting past, current or future tasks or actions	Sharing, spreading of knowledge about work content and work methods.	Discussions and negotiations concerning: work, trade or organization regulations, paradoxical orders or impediments. Reminder of rules and constraints. Suggestions for alternatives to deal with them	Propositions and discussions on the new modes of functioning or new criteria and rules concerning the realization of work	Dialogue and confrontation of points of view and advice on the quality of work	Exchanges and discussions concerning "significant" changes in joint activity (modification of the objective of the activity, its goals, new functions, new missions, new projects

Table 14.2. *Grid for observing collective activity (Crouzat 2019)*

XXX: appears in 100 of the uses observed XX: appears in more than 50% but less than 100% of the uses observed X: appears in the uses observed			Collective activity resources mobilized							
			COLLECTIVE WORK				WORK COLLECTIVE			
			Operative synchronization	Cognitive synchronization						
Tools observed	Number of participants observed	Duration of observations (in hours)		Concerning the current situation	Concerning understanding and mutual visibility	Through the sharing of knowledge	Conflict and regulation management in action	Development and/or re-development of work and of the profession	Debate on the quality of work	Contribution to the common history of the activity
E-mails	65	25	X	XX	X	XX	XX		X	X
Diary manager	43	4	XXX		X					
Video conferencing system	25	5.5	X	XXX	XX	X	X			
Instant messaging	2	2	X	X	X	XX	XX		XX	
Shared files	57	12	X	X	X	XXX				X
Corporate social network	502	20		X	XX	XX	X		X	X
PowerPoint software	326	25		XX	XX	X				

Table 14.3. *Breakdown of collective uses according to the tools observed and the collective activity resources mobilized (Crouzat 2019). For a color version of this table, see www.iste.co.uk/bobillierchaumon/digital.zip*

14.5. Main results: virtually hyper-instrumented collective work for invisible work collectives

Table 14.3 presents the results from the observations made using our grid. It highlights the components of collective activity mobilized during the use of digital collaborative artifacts.

We note that the uses of the tools are mainly for synchronization (both operative and cognitive) of collective work resources: distribution of appointments and tasks, visibility of work in progress, adjustments, indicators, etc. Few uses ultimately allowed the faculties of expression necessary for the development of work collectives that may have been in place. In fact, only the use of e-mail and video conferencing generated discussions on management of conflicts and rules in action, a debate on the quality of the work or a contribution to the common history of the activity (via, for example, the proposal of new projects or the spontaneous creation of informal work groups). E-mail, like video conferencing, allowed for less formal expression (free writing, oral). These tools had the necessary flexibility to allow actors to express points of view, to discuss them through interactive messages (or directly within the framework of the video conference) and therefore to go beyond the simple exchange of information. Both were more conducive to expression and debate or controversy. This was less the case with tools such as the diary manager, file sharing (purely informative media) or PowerPoint software (used to illustrate oral presentations).

We noted that the effects of digital collaborative tools were directly dependent on actual use:

– the collective work of articulation and synchronization of tasks and missions was largely supported by the use of these tools as memos, trackers, reminders and support for information sharing (e.g. a PowerPoint presentation, a medium for information presented orally);

– the work collective's resources were instead sought by means of tools that allowed expression, spontaneity and criticism of work in progress or to come.

In addition to these results from our observations, the relatively long usage and completion times of these tools (intelligence) often tended to replace more spontaneous and informal "free" dialogue.

14.6. Discussion

Digital tools facilitate and support the synchronization of individual activities by enabling coordination, shared visibility of the ongoing activity and distribution of tasks. In this sense, these tools can promote collective work. They respond to a need for traceability of the progress made in the work and are mainly used as pure informative media, and less in expression. Nevertheless, the work collectives studied did not benefit from the use of these tools specifically. They did not allow for dialogue, debate or discussion of shared rules. Nor did they allow for the sharing or discussion of concrete professional action; they did not specifically promote the consolidation of professional types and their evolution because they did not directly facilitate free, critical and spontaneous debate concerning the quality of the work in progress. They remained channels for debate but were rarely used in this sense; they presented few opportunities for expression and debate on the quality of work or its evolution over time (conditions, etc.). These tools were essentially used as simple informative media with no inention of creating links or interchanging spontaneous ideas.

At the same time, the use of these artifacts (time to devote to them) took precedence over most of the time for collective reflection, confrontation and debate (spontaneity). In addition to information saturation, there was also a saturation of digital actions to be performed to process/arrange/sort the information constantly transmitted digitally. This last activity (often individual) developed with the number of different media used (which could sometimes change according to projects, functions and teams). There were too many management and communication tools, which seemed to fragment the collective work: they multiplied the common operational reference systems in force (by increasing the individual work of completing and informing the tools). Too formalized, the tools, digital or not, also hindered the opportunities for expression and debate, which were nevertheless favorable to the development of collective work.

With respect to the quality of life at work, it was also likely to be impacted by these so-called collaborative digital tools: there was a risk of weariness and frustration in promoting only "transmitter–transmitter" type activities to the detriment of the actions of co-construction, listening and mutual criticism, reflexive trial and error and debate, all necessary for any reliable collective production. Our other studies carried out in parallel (Crouzat 2019) also highlight the frustration of the actors, which they associated with the slowing down of what they called "real work".

The digitally mediated collaboration did not therefore pose a real problem in itself. The media were even likely to significantly improve collective work. In particular, they could promote remote exchanges and increase the speed of these exchanges. They facilitated the interpretation of information and enabled simple storage. They supported the synchronization of tasks and missions. These qualities were well known and employees confirmed them in other studies we have conducted (Crouzat 2019). However, the over-exposure of the activity, as well as too many digital tools, led to more individual activity and many adjustments to use and complement them. This mechanism explains how too great a duplication of digital tools contributes to the fragmentation and dispersion of shared operative repositories (duplication of information in different formats), and how the multiplication of so-called collaborative tools can sometimes increase individual work (information capture) to the detriment of the collective work itself.

14.7. Conclusion

It is indeed the type of media used and their conditions of use that can have a negative impact on collective activity: by hindering the resources of work collectives, that is, by blocking the free development of the activity. These types of uses are closely related to the rules, habits and organization of work in place: the switch, the absence of "emptiness" (absence of non-executive moments, moments with no process to be applied), speed and use in an emergency. Finally, the format of the medium, associated with the type of use, also plays its role in the place it leaves for the subject (expression, precision, explanations). Technology must not hinder the mobilization of subjectivity at work (Bobillier Chaumon 2018) by prescribing too much rationality.

The technology is based on a binary and coded language that does not yet take into account all specifically human capacities for expression and criticism (Crouzat 2019). Human communication and exchanges at work are not limited to the exchange of information and mutual adjustments. They require a certain amount of expression, creativity and criticism that is limited by certain tools when they are too prescriptive, too formalized and/or used solely to perform a simplification function. Once again, this observation is in line with the work of Bobillier Chaumon (2018): flexibility is a guarantee of optimal and efficient use of digital technology in the service of work (and therefore health). Indeed, if we value – too highly – a reduction in time spent on the specifically human activities of deciphering, appropriation and interpretation, then the formal activities of encryption and intelligence go on multiplying, sometimes at the expense of reflection, discussion and creativity – uniquely human needs ever present amongst employees (which technology cannot reproduce).

14.8. References

Bobillier Chaumon, M.-E. (2013). Conditions d'usage et facteurs d'acceptation des technologies dans l'activité : questions et perspectives pour la psychologie du travail. HDR Thesis, University of Grenoble 2, Grenoble.

Bobillie Chaumon, M-E. (2017a). Du rôle des TIC dans la transformation digitale de l'activité et sur la santé au travail. *La Revue des conditions de travail de l'ANACT*, (6), 16–24.

Bobillier Chaumon, M.-E. (2017b). Transformation numérique et charge de travail : comprendre les liens. In *Les Cahiers des RPS*, Martini, F. (ed.). Ministère du travail, Paris.

Bobillier Chaumon, M.E. (2018). L'expérience utilisateur dans l'appropriation des technologies : quelles approches possibles ? *Psychologie du travail et des organisations*, 24(4), 309–312.

Bobillier Chaumon, M.-E. and Clot, Y. (2016). Clinique de l'usage. Les artefacts technologiques comme développement de l'activité. *Activités*, 13(2).

Caroly, S. (2010). Activité collective et la réélaboration des règles : des enjeux pour la santé au travail. Thesis, Bordeaux Segalen University, Bordeaux.

Caroly, S. (2011). Activité collective et réélaboration des règles comme ressources pour la santé psychique : le cas de la police nationale. *Le travail humain*, 74(4), 365–389.

Caroly, S. (2012). Groupes et collectifs de travail. In *Manuel de psychologie de travail et des organisations : les enjeux psychologiques du travail*, Sarnin P. and Bobillier Chaumon M.E., (eds). De Boeck, Paris.

Chiapello, E. and Gilbert, C. (2012). Les outils de gestion : producteurs ou régulateurs de la violence psychique au travail. *Le travail humain*, 75(1), 1–18.

Caroly, S. and Barcellini, F. (2013). Le développement de l'activité collective. In *Ergonomie constructive*, Falzon P. (ed.). PUF, Paris.

Clot, Y. (2000). La fonction psychologique du collectif. In *Le travail collectif. Perspectives actuelles en ergonomie*, Weill-Fassin, A., and Benchekroun, T.H. (eds). Octarès, Toulouse.

Clot, Y. (2008a). *Le travail sans l'Homme ? Pour une psychologie des milieux de travail et de vie*. La Découverte, Paris.

Clot, Y. (2008b). *Travail et pouvoir d'agir*. Presses Universitaires de France, Paris.

Clot, Y. (2010). *Le travail à cœur. Pour en finir avec les risques psychosociaux*. La Découverte, Paris.

Crouzat (2019). Collectifs de travail en ingénierie aéronautique : comment soutenir et développer l'activité collective en contexte de transformation organisationnelle et digitale? PhD Thesis, GRePS, Lyon 2 University, Lyon.

Crouzat, P. and Bobillier Chaumon, M. (2017). Technologies collaboratives : usages et impacts au sein de l'activité collective. *La revue des conditions de travail*, 145–152.

Crouzat, P., and Bobillier Chaumon, M.-E. (2018). Recherche-intervention : acceptation située et instauration d'une activité dialogique dans l'élaboration d'un collectif de travail. In *Bien-être et diversité des situations de travail*, Daele, A.V., Hellemans, C., and Casini, A. (eds). L'Harmattan, Paris.

Dejours, C. (1993). *Travail, usure mentale : de la psychopathologie à la psychodynamique du travail*, Bayard Éditions, Paris.

de la Garza, C. and Weill-Fassina, A. (2000). Régulations horizontales et verticales du risque. In *Le travail collectif*, Benchekroun, T.-H. and Weill-Fassina A. (eds). Octarès, Toulouse.

Grosjean, M. and Lacoste, M. (1999). *Communication et intelligence collective : travail à l'hôpital*. PUF, Paris.

Leplat, J. (1991). *L'analyse du travail en psychologie ergonomique : recueil de textes*. Octarès, Toulouse.

Leplat, J. (2001). *Les compétences en ergonomie*. Octarès, Toulouse.

Litim, M. and Kostulski, K. (2006). Le diagnostic d'une activité complexe en gériatrie. *Nouvelle revue de psychosociologie*, 45–54.

Poret, C. (2015). Concevoir pour le pouvoir d'agir ensemble d'un collectif transverse : le cas de la relation de service dans le domaine commercial. University Paris 8, Paris.

de Terssac, G. and Chabaud, C. (1990). Référentiel opératif commun et fiabilité. In *Les Facteurs humains de la fiabilité dans les systèmes complexes*, Leplat, J. and de Terssac, G. (eds). Octarès, Toulouse.

Wisner, A. (1993). L'émergence de la dimension collective du travail. *Actes du 27e Congrès de la SELF*. Octarès, Toulouse.

PART 4

Approaches and Methods for Conducting Digital Transformations

15

Prospective Ergonomics in Service of Technological Innovation

The purpose of this chapter is to explain the place and role of prospective ergonomics (PE) in the development of future products, services, processes and systems and to present two illustrations of innovative projects applying its approach and methods. First, we will define and characterize PE, explain the context of its creation and development and then present the challenges that need to be overcome in order to meet the new expectations it creates among researchers and practitioners.

15.1. Introduction

Ergonomics has long been kept out of the strategic decisions of organizations, because of the types of interventions taking place and the mission it had set for itself. In fact, it has essentially been used to correct problematic human work situations – often playing the role of a firefighter – or to participate in the design of tools, workstations, systems and work environments used by humans – in this case, playing a role in supporting design disciplines. This has been beneficial for it, judging by the vast body of knowledge, methods, tools and standards that have been developed over the years and the recognition it receives in the workplace. Ergonomics, on the other hand, cannot really make a positive assessment of its own inventions, novel ideas or successes of products, services, processes or technologies

---

Chapter written by Jean-Marc ROBERT, Antoine MARTIN, Mitra TARAGHI, Clément COLIN, Masood MALDAR, Flavie BONNEVIOT and Eric BRANGIER.

that are its own initiative. This is hardly surprising because it has not positioned itself in this way and is not yet recognized as a discipline of creation and innovation.

But times are changing and ergonomics is constantly evolving to remain relevant, modern and capable of producing economic and human value in today's scientific, technological, economic and social contexts. Thus, its service offer and scope of action are expanding with the creation of prospective ergonomics (PE), a new type of intervention focused on exploring current and future needs, discovery, innovation and the creation of new artifacts, services and systems.

The aim of this chapter is to explain the place and role of PE in the development of future products, services, processes and systems, by looking at two illustrations of innovative projects that are applying its approach and methods. First, we will define and characterize PE, explain the context of its creation and development and then present the challenges that need to be overcome in order to meet the new expectations it creates among researchers and practitioners.

15.2. A new form of intervention focused on creation and innovation

Prospective ergonomics is a new modality of ergonomic intervention focused on the creation of future products, services, systems and processes (Brangier and Robert 2014). It differs from corrective ergonomics and design ergonomics. The former aims to correct existing tools, machines, systems and work situations to make them safer, more efficient and capable of providing good user experience, while design ergonomics is involved in the design of new artifacts for human use, although these artifacts are often already identified and largely predefined by other professions. Table 15.1 illustrates four points on which the ergonomics of correction, design and prospecting differ from one another (ability to act, actions and objectives, involvement in technological development, temporality). However, this should not obscure the fact that the three modalities of intervention also have a great deal in common and can be linked to each other in practice since:

– these three modalities may require creativity and lead to technological discoveries and innovations; therefore, PE does not have a monopoly in these matters, although it is more naturally associated with it;

– they focus on what is the hallmark of ergonomics: human performance and well-being (in terms of health, safety and user experience);

– in all three modalities, a minimum amount of prospecting and anticipation is required about users, activities, needs, uses, contexts, constraints, ... in order to propose corrective measures or to design artifacts that can survive over time;

– each type of intervention can be relayed by another type of intervention; for example, corrective ergonomics will be followed by design ergonomics if a new artifact must be developed to correct the situation; conversely, design ergonomics will be followed by corrective ergonomics if problems that were not detected during the design phase must be corrected; PE that intervenes upstream of projects will be relayed by design ergonomics once the idea of the new artifact that is to be designed is fairly well defined;

– finally, the same methods of data collection can be found in all three modalities: for example, interviewing, observation, reasoning aloud, communication analysis, taking measurements, task and scenario analysis.

	Prospection	Design	Correction
Capacity for action	High	Average	Low
Actions and objectives	Imagining and producing future scenarios to amplify the performance, well-being and health associated with the development of future products, services, stations and systems.	Implementing ergonomic knowledge to design new artifacts, with a view to maximum performance, well-being and satisfaction.	Implementing ergonomic knowledge to correct existing artifacts, with a view to maximum performance, well-being and satisfaction.
Technological development	Innovation focused	Development focused	Redevelopment focused
Project time frame	Future Imagination and description of non-existent possibilities.	Present Realization of a possibility and preparation of the use.	Past Correction, improvement or transformation of the existing.

Table 15.1. *Some elements that distinguish the three ergonomic intervention modalities*

15.3. The context and challenges of prospective ergonomics

PE intervenes in a difficult context when everything has to be defined: there are no precise or initial requests from a client about any problem that need to be corrected or an artifact that needs to be developed, so there is no team or budget allocated. The future needs to be met, future users (some may not yet be born), future human activities, future tools and systems that will be part of the technological ecosystem, future artifacts that need to be developed whose future context of use are unknown. To succeed in such a work environment, the ergonomist must have a certain state of mind and combine several qualities: he or she must remain on the lookout for emerging or future needs and the positioning of competitors, be visionary, proactive (to initiate actions), intuitive (to seize business opportunities), creative (to imagine new, relevant and innovative solutions), convincing (with decision-makers and investors) tolerant of risk (inevitable in innovation and for future artifacts) and be a leader (to train and lead a team of collaborators).

PE intervenes in three types of contexts (Table 15.2) corresponding either to a technology driver, a market reader or a need seeker (Jaruselski *et al.* 2014). The first (technology driver) is marked by the advancement and preponderance of technology, for which useful and new applications are sought in relation to needs that will be discovered, or human activities to be supported. The second (market reader) is marked by known market needs, which then serve as a driving force for technological development or the search for new ways of organizing our activities. The third (need seeker) is marked by the exploration of needs that will be discovered which, as in the previous case, will be at the origin of new technological developments or new forms of organization for our activities. In these three contexts (and especially the third), the analysis of needs (of clients, consumers or users) is central, hence the need to agree on the characteristics of the needs.

Consistent with Griffin (2013), we believe that needs are fairly stable, changing only slowly, if at all, over time. Therefore, strictly speaking, it would be inappropriate to talk about future needs because for all intents and purposes, they are the same as current needs. Despite this, we will continue to talk about *future needs* because, in addition to being evocative, it is a term commonly used in literature on the development of new artifacts, including prospective ergonomics. What is rapidly evolving are the ways of meeting the needs, in other words the proposed solutions, and the levels that need to be achieved on a set of evaluation criteria: faster, safer, more economical, more ergonomic, more inclusive (for people with disabilities), less energy consuming, more respectful of the environment (favoring circular economy and sustainable development), can be done remotely, more in line with the new values of society (e.g. purchase of local and fair trade products), more complete, more technologically integrated, more comfortable, more esthetic, etc. This reality affects all areas of activity, such as food, housing, clothing, transportation, work, education, leisure, services, etc. In each field and in the three contexts defined above, imagining the future in terms of needs, artifacts, contexts, human activities and interactions is at the heart of prospective ergonomics. It must therefore enrich its own knowledge, approaches and methods, and collaborate with other disciplines in the development of new products to meet the challenge of the future, which is often characterized by a high level of complexity, uncertainty and ambiguity.

Context	Technology driver	Market reader	Need seeker
Objective	Develop technologies that will create new uses.	Use market reader to find value-creating ideas to develop new artifacts, new technologies and new forms of organization.	Research current and future needs in order to find ideas for value creation, to develop new artifacts, new technologies and new forms of organization.
Focus	**The technology** Technology application, science and engineering.	**The market** The emergence of a promising market segment. The emergence of a social demand.	**The need** The anticipation of a new use. The construction of a new form of satisfaction.

Table 15.2. *The three contexts of innovation*

In addition to the requirements of the three innovation contexts that push ergonomics to adapt and develop different intervention approaches, there are two major issues that affect the development and recognition of the discipline: the search for new knowledge, approaches and methods to work with the future, and the positioning of PE at the strategic level of organizations.

First, PE's approach to broadening and enriching ergonomics calls for similar action on the knowledge side. In dealing with new problems and demands, ergonomics researchers and practitioners are naturally led to broaden their fields of reflection through the active search for new knowledge and the implementation of new approaches and methods. This is so they will be able to respond appropriately to the new innovation challenges they face. EP is broadening the scope and reach of ergonomics by assigning new roles to ergonomists: acting on their own initiative to initiate, define and direct projects, intervening upstream of projects to discover the needs, problems and requirements that will guide design, and relying on creativity, discovery and innovation in framing problems and finding solutions.

Second, PE is clearly positioned at the strategic level of organizations, where ergonomics is rather absent (Dul and Neumann 2009). For any development of new products, services or large-scale systems, the deployment strategy must be in line with the organization's mission, objectives, development plan, priorities and human, technical and financial resources for optimal profitability. Accordingly, the development of innovative and competitive artifacts generally requires such a large amount of resources (human, technical and financial), that decisions must be made at the strategic level of the organization to anticipate costs, needs, timelines and risks upstream of the project. As PE is involved both upstream and during project creation, the ergonomist as an internal member of the organization, will have no choice but to seek the support of senior management. For its part, senior management may see the advantage of having one more unit within the organization that focuses on creativity, discovery and innovation and, in doing so, helps it to become more competitive.

15.4. Foundations in several disciplines and fields of study

The theoretical and methodological foundations of PE come from two-parent disciplines (ergonomics and prospective) and two fields of study from psychology (creativity and future-oriented cognition). These are outlined in the following paragraphs.

– **Ergonomics**. As a subdomain of ergonomics, PE naturally inherits all the knowledge, approaches, methods and standards of ergonomics, including those of closely related domains such as human–computer interaction, interaction design, usability, user experience design, to study human–human and human–artifact interactions (including intelligent agents) for the purpose of innovation in products, services, processes and work systems. For example, ergonomic task analysis (Hollnagel 2012) and users' journey map analysis, which are well-known methods in ergonomics and human–computer interaction, can be very useful in PE to understand the activities, experiences and emotions related to the use of different artifacts and to identify elements that could be innovative.

– **Prospective**. PE needs the insights of prospective, a discipline whose role is to gather information and conduct structured reflection on the future, in order to guide decisions and actions. Prospective enables the construction of scenarios of likely future situations involving contexts, people, activities, artifacts and uses that help explore, explain and shape future change. It is not about trying to predict the future (unlike modeling, another branch of anticipation, distinct from prospective), which would be futile, but about imagining and preparing for it in the light of information about trends and what they predict. As Ratcliffe (2006) argues, in order to create the future, we must first be able to imagine it. Several well-known methods are used in prospective to obtain information on trends and build possible futures, including technology monitoring, trend analysis (e.g. demographics, lifestyles, consumption, labor, values) and the scenario method.

– **Creativity**. Creativity, which is the basis of discovery and innovation, which are in turn, the basis of wealth creation, is essential to PE throughout the artifact development process, especially when it comes to seeking new and adapted solutions to meet needs. There is an abundant literature on creativity from research in psychology, design and marketing (Bonnardel 2006; Sternberg and Kaufman 2018). It offers us a panoply of methods to stimulate the production of ideas (e.g. brainstorming, analogy, role-play, improbable encounter technique, logbook) as well as advice on how to use each method in order to obtain better results. PE can benefit directly from this important body of knowledge.

– **Future-oriented cognition**. This type of cognition (D'Argembeau *et al.* 2011; Osvath and Martin-Ordas 2014) refers to a person's ability to mentally project himself or herself into the future, to consider a future experience by mentally creating a realistic scenario. This is a subject of study that is rich in questions that can be investigated, when we know that humans have difficulty imagining the future or mentally projecting themselves into it. Although this subject of study is still poorly documented, poorly known and is not considered at the time of the creation of PE, it represents a great potential for the development of PE, considering its strong face validity and the results of early methodological work (Allinc et *al.* 2018; Colin and Martin 2019). Research on future-oriented cognition has no shortage of subjects to investigate, including: characterizing individuals who are more successful at projecting themselves into the future; comparing the results of projections from different individuals or groups; knowing the ways of doing things, the cognitive processes involved, the levels of perceived effort and the levels of confidence in our own projections; testing the impact of different beginnings on projections and assessing the accuracy of projections over time.

15.5. Prospective ergonomics intervention approach

Figure 15.1 shows an example of a PE intervention approach upstream of the design process. This approach consists of six phases with associated feedback loops and iterations that show that the results of one phase can lead us to revisit information and decisions from earlier phases and reposition ourselves. Needs analysis benefits from learning from the feedback of those who are asked to use and respond to numerical simulations of artifacts, mock-ups and prototypes. Such artifacts and prospective scenarios help to refine, supplement and prioritize needs.

Phase 1: *Use prospective data*. According to our interests, our knowledge, our means of action and/or our intuitions, choose a field of activity, a subject, a question, a problem to be dealt with and a development opportunity to be seized. This can be in any field of activity: for example, passengers air transport, food wastage, international student exchanges, labor scarcity, use of autonomous cars, distance technical online training, etc. For the chosen topic, collect prospective data and recognize the state of the current situation (e.g. problems to be solved, customer expectations, artifacts in use, out of step with technological progress) and the opportunities for development or change through new technologies, or new forms of organization of activities. It is also about demonstrating the importance of the future for organizations and/or society and exploiting the results of prospective studies and statistical data from prospective resource centers (e.g. *Futuribles*, *Institute for the Future*, *National Intelligence Council*).

Phase 2: *Understand future developments*. Produce and analyze data on typical and atypical users; future activities; possible reference situations; lived, desired and imagined experiences; artifacts used; problems encountered; sources of satisfaction and dissatisfaction; opportunities for improvement or radical change. Detect weak signals. Use foresight to agree on likely future events, identify trends, show the extent of anticipated changes and assess impacts.

Phase 3: *Describe prospective scenarios.* Process the data to extract representations of future users, activities, contexts, artifacts and uses. Use frameworks and models to illustrate the artifacts that are being considered, the personas method to define the characteristics of the intended future users and the scenario method to describe future needs, activities, interactions with artifacts and use context.

Phase 4: *Stimulate ideations*. Stimulate and amplify the creativity of people (e.g. experts, creatives, extraordinary user leaders) who participate in defining possible future scenarios and the future artifacts envisioned, with participatory and creative techniques.

Phase 5: *Ratify prospective scenarios*. Criticize, prioritize and confirm ideas about future users, needs, artifacts, activities and usage contexts. Move ideas towards future products, services, systems or processes.

Phase 6: *Help with decision-making*. Select one or more design projects in light of initial estimates made on the basis of a set of criteria: added value compared to the existing one, feasibility, costs, required means of development, ease of insertion into the future technological ecosystem, situation of competitors, market potential, ethical quality and social acceptability.

The outcome of this intervention approach by PE is the proposal of one or more innovation projects for the future, which will then go into the center of design ergonomics, which can then rely on a set of approaches, methods, tools and standards that ensure good results. Ergonomic design advocates a user-centered design approach (ISO 9241) whose main activities are as follows:

– recognize the users, the tasks and the context;

– build a prototype of the system to be tested;

– make end-users perform realistic tasks;

– take measures of performance and human satisfaction;

– follow an iterative approach.

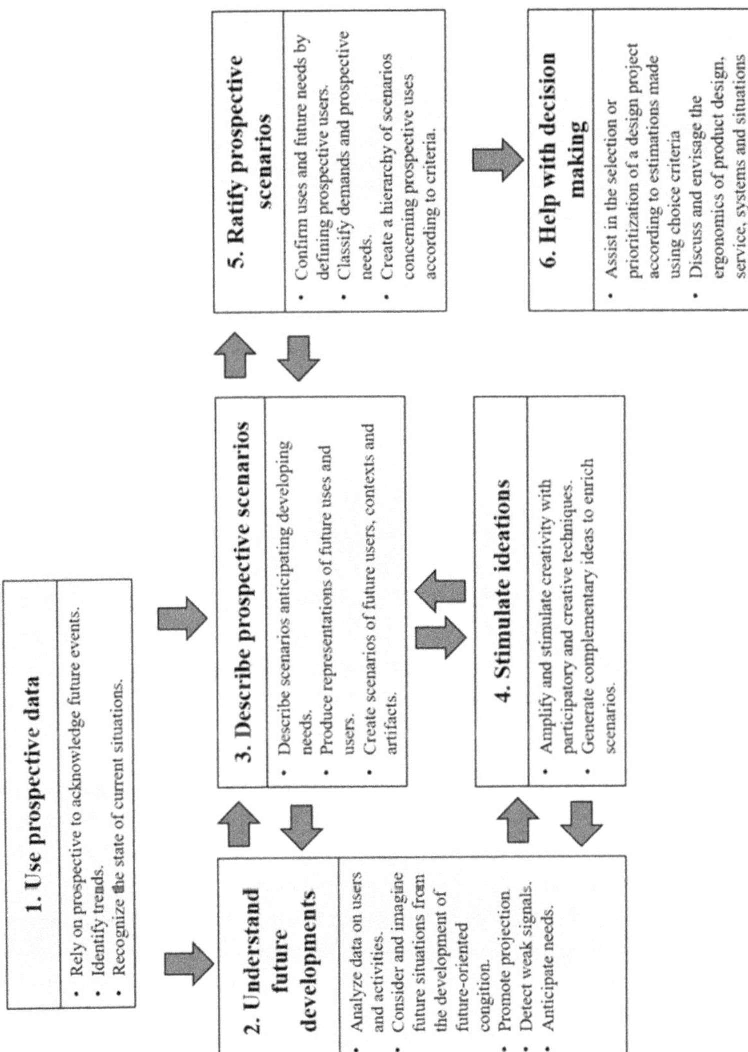

Figure 15.1. *The six phases of a prospective ergonomic intervention approach*

The user-centered design approach, through context analysis, is not limited to the end-user, but also looks at other parties that may be involved in the design process: clients, supervisors, developers, trainers, technical support, customer service, etc. Each party is likely to enrich the design project by proposing functionalities, requirements to be met, evaluation criteria, ways of doing things and qualities to be met. Some ergonomists go further than the user-centered design approach, by favoring a participatory design that allows the various parties to participate in decision-making.

15.6. Two cases of intervention

15.6.1. *Anticipating future needs and uses related to energy systems that integrate hydrogen energy*

Hydrogen energy is a new energy carrier that has been identified to have key potential for energy transition (Cany *et al.* 2017). It should enable the development of renewable energies by allowing their storage, and making energy production (fluctuating with renewable energies) and user demand coincide. The housing sector is the second largest consumer of final energy in France, with more than 60% using carbon energy. In this context, hydrogen energy is seen as a means of contributing to the energy transition of the residential sector, in particular by allowing the consumption of renewable energies that could possibly be produced in the building sector. Energy systems that integrate hydrogen energy for housing are currently mostly experimental. Thus, the prospective ergonomics approach is deployed here in a framework for which the context is poorly defined and for a rather distant time frame.

The objective of this intervention was to anticipate future needs and uses related to residential energy systems that integrate hydrogen energy. To do so, a three-step approach was used. First, 36 interviews to explore future needs and uses (Brangier *et al.* 2018) were conducted with energy and housing experts, prospective users – who had precursor uses – and ordinary users. The purpose of these interviews was to gather participants' knowledge of the needs and uses of future energy systems. Participants verbalized their experience of energy for habitat on the basis of activity maps constructed from an analysis of energy-related activity. They were then led to imagine the future of these activities through methods of future mental travel. The data collected during the interviews was analyzed, using various lexical and content analysis techniques, in order to generate innovative ideas that could lead to new systems, products and services. A total of 231 ideas were identified. In the second step, this data was used to stimulate ideas from energy experts. Finally, the ideas evaluated as most relevant were selected to establish a panorama of possible innovations.

In this work, the ergonomist intervened upstream of technological development and allowed the design team to rely on a description of future users and future use contexts, in order to generate future use scenarios that corresponded to product and service ideas. Thus, the objective was to guide the development of hydrogen technologies upstream, by considering experts and users as resources that promoted design.

15.6.2. *The crowdsourcing approach*

This method of data collection is increasingly used to involve large numbers of people in ideation, design and production tasks, hence its interest in PE (Robert *et al.* 2019). Among the 22 definitions of crowdsourcing that we counted, we retain this one, which covers the important aspects on which the authors seem to agree: "The fact that a company or institution takes on a function formerly performed by employees and subcontracts it to an indefinite and generally extensive network of people, in the form of an open call" (Howe 2006). Two requirements must be met: an open call and a wide network of potential workers.

This participatory production method has been shown to be capable of producing successful results for a variety of tasks, ranging from small and tedious, to larger tasks involving complex problem solving and innovation. Indeed, it is part of two major currents in the field of innovation: user-driven innovation (von Hippel 2013) and open innovation (Aitamurto *et al.* 2011).

The essential components of participatory production are, on one hand, individuals who come from the crowd and form the workforce, on the other hand, companies that tender services to people outside of their walls to perform various tasks, and finally, an online tool based on social media, that connects the crowd and the companies to each other (Schenk and Guittard 2011). Thus, depending on the skills required by the project, a very large number of people have the opportunity to participate in the proposed project, which contributes to the democratization of innovation. The method allows the company to have access to new and varied talents from outside of the company, who participate voluntarily and sometimes even without remuneration, in the resolution of a problem, whatever the field. Hippel (2013) argues that this willingness to work for free (in many cases) can be explained by the fact that the participants are intrinsically interested in the task or challenge at hand, that they value their innovation process and that they can take pride in contributing to something that makes sense to them.

The method has been used to support different types of activities: the production and selection of ideas, competition for innovation, accomplishment of small tedious tasks or larger tasks, and creation of common good for a non-profit organization (e.g. Wikipedia). It is supported by various software.

15.7. Conclusion

PE introduces the question of the ergonomist initiating innovation projects through his or her activities of anticipating needs, upstream user research and imagining possible future scenarios. It also introduces the search for creativity and innovation in the creation of new artifacts or new forms of organization. This requires a vision of the development to be carried out and intuition. PE is thus at the heart of business processes and strategic orientations because it aims to develop new artifacts that generally require many resources. Through its human-centered innovations, it has the capacity to create social and economic value, and it opens up a whole new field of action for the ergonomist.

Concerning the methodological aspect (Robert and Brangier 2019), the question arises as to the positioning of PE approaches and methods in relation to other future-oriented user-centered approaches (e.g. design fiction, needfinding, speculative design, strategic design, design thinking, etc.). In order for the ergonomist practitioner to be able to find his or her way around, research is necessary to test, adapt if necessary and validate, in a *prospective* context, the methods derived from design ergonomics, related fields (human–computer interaction, usability, user experience), design disciplines (design, engineering) and marketing, in order to know which ones give the best results. Will there ultimately be original approaches and methods from PE? Moreover, it is necessary to test and adapt methods coming from PE, such as the scenario method, the Delphi method, futures wheel and so on, which have not yet been deployed in prospective ergonomics projects.

Concerning the theoretical part, scientific knowledge on future-oriented cognition is still very limited. Their development needs to be accelerated because it is likely to have the power to form a solid basis for PE. Several questions arise about it: what cognitive capacities do humans rely on to project themselves into the future of their activities? Could these capacities be refined to improve the quality of projection data? How can they be stimulated and strengthened? Using what type of initial element? How can we detect upstream users with better capacities to project themselves into the future than others? Could we model this propensity to develop prospective expertise, with the aim of making them priority users in PE approaches? These are avenues of research on human cognition that should strengthen the development of PE.

15.8. References

Aitamurto, T., Leiponen, A., and Tee, R. (2011). The promise of idea crowdsourcing – Benefits, contexts, limitations. *Nokia Ideasproject White Paper*, 1, 1–30.

Allinc, A., Cahour, B., and Burkhardt, J.M. (2018). The guided imaginary projection, A new methodology for prospective ergonomics. *Proceedings of the 20th Congress of the International Ergonomics Association*. IEA 2018. Springer, Cham, 1340–1347 [Online]. Available at: www.doi.org/10.1145/3317326.3317340 [Accessed 10 November 2019].

Bonnardel, N. (2006). *Créativité et conception, approches cognitives et ergonomiques*. Groupe de Boeck, Paris, France.

Brangier, É. and Robert, J.-M. (2014). L'ergonomie prospective : fondements et enjeux. *Le travail humain*, 77(1), 1–20.

Brangier, É., Brangier, B., Marache-Francisco, C., Kopp, S., and Clausse, J. (2018). Un processus d'interview pour produire une expertise sur l'anticipation des besoins futurs. *Proceedings of the 16th Ergo'IA "Ergonomie et Informatique Avancée" Conference*. Ergo'IA '18 [Online]. Available at: www.doi.org/10.1145/3317326.3317340 [Accessed 10 November 2019].

Cany, C., Mansilla, C., da Costa, P., and Mathonnière, G. (2017). Adapting the French nuclear fleet to integrate variable renewable energies via the production of hydrogen: Towards massive production of low carbon hydrogen? *International Journal of Hydrogen Energy*, 42(19), 13339–13356.

Colin, C. and Martin, A. (2019). Improving mental time travel abilities to support the collection of needs: A pilot study. *Proceedings of the 50th Nordic Ergonomics and Human Factors Society Conference*, Broberg, O. and R. Seim (eds). Nordic Ergonomics and Human Factors Society, Elsinore, Denmark [Online]. Available at: hal.archives-ouvertes.fr/hal-02348430 [Accessed 10 November 2019].

D'Argembeau, A., Renaud, O., and Van der Linden, M. (2011). Frequency, characteristics and functions of future-oriented thoughts in daily life. *Applied Cognitive Psychology*, 25(1), 96–103.

Dul, J. and Neumann, W.P. (2009). Ergonomics contributions to company strategies. *Applied Ergonomics*, 40(4), 745–752.

von Hippel, E. (2013). User innovation. In *Leading Open Innovation*. MIT Press, Cambridge.

Howe, J. (2006). The rise of crowdsourcing. 1 June 2006 [Online]. Available at: https://www.wired.com/2006/06/crowds/ [Accessed 21 November 2019].

ISO DIS (2010). 9241-210: 2010. Ergonomics of human system interaction. Part 210: Human-centred design for interactive systems. International Standardization Organization (ISO), Switzerland [Online]. Available at: www.iso.org/cms/render/live/en/sites/isoorg/contents/data/standard/05/20/52075.html [Accessed 10 November 2019].

Jaruselski, B., Staack, V., and Goehle, B. (2014). The global innovation 1000: Proven paths to innovation success. *Strategy+Business*, 77, 2–16.

Osvath, M. and Martin-Ordas, G. (2014). The future of future-oriented cognition in non-humans: Theory and the empirical case of the great apes. *Philosophical Transactions of the Royal Society B*, 369(20130486).

Robert, J.-M. and Brangier, E. (2019). Overview of methods for prospective ergonomics. In *Proceedings of the 50th Nordic Ergonomics and Human Factors Society Conference*. Broberg, O. and R. Seim (eds). Nordic Ergonomics and Human Factors Society, Denmark.

Robert, J.-M., Maldar, M., Taraghi, M., and Seffah, A. (2019). User innovation, Lead users and crowdsourcing for the design of new products and services: Why, What and how? In *Proceedings of the 20th Congress of the International Ergonomics Association (IEA 2018): Advances in Intelligent Systems and Computing*, Bagnara, S., Tartaglia, R., Albolino, S., Alexander, T., Fujita, Y. (eds). Springer International Publishing, Cham, Switzerland.

Schenk, E. and Guittard, C. (2011). Towards a characterization of crowdsourcing practices. *Journal of Innovation Economics and Management*, 7(1), 93–107.

Sternberg, R.J. and Kaufman, J.C. (eds) (2018). *The Nature of Human Creativity*. Cambridge University Press, Cambridge.

16

Simulating Digital Activity in the Making: Elements of Methodology

Digital transitions are not limited to technological transformations, but embody with them organizational, spatial, social, material transformations and so on. This chapter describes the interest of activity simulation, to globally apprehend the new work configurations that prefigure these transformations, and help to design them. Through two illustrations, it shows in particular how digital activity, considered as a human activity developed within a digital economy, is characterized by very concrete dimensions that simulation devices must be able to account for, in order to help the subjects keep control of it.

16.1. Introduction

The "digital tsunami" (Valenduc and Vendramin 2016) that some media are announcing should, if we are to believe them, sweep away the old worlds of economy, employment and work, replacing them with a now digital economy, a predominantly uberized job and an inevitably virtual job. Faced with this prospect, it is enough to make us dizzy, as it seems difficult to imagine the outlines of tomorrow's work. How can we plan for it? How can we prepare for it? How can we act on it and put it in our hands rather than the other way round?

Notwithstanding the fact that the announced disruptions are not always considered as such and that the future transformations are in part an acceleration of existing trends (ibidem), it can also be argued that tomorrow's work activities will always have very concrete dimensions, that is, located in the field of the sensitive experience of a subject interacting (physiologically, cognitively and subjectively)

Chapter written by Laurent VAN BELLEGHEM.

with an equally concrete environment (made up of spaces, technical systems, systems of rules, social interactions, etc.). "Digital activity" is therefore not an abstract activity (as the notion of "virtual work" might suggest) but on the contrary, is quite concrete, embodied and experiential. From then on, we can plan for it and even act, in terms of design, on the characteristics of the situations that give it life. Activity simulation (Van Belleghem 2018) should be able to help. In this context, it would be useful to consider simulation devices that facilitate the *concrete* understanding of changes brought about by digital technologies. The development of *tangible*, rather than digital devices, is a counter-intuitive approach that can, and indeed must, be defended.

16.2. Digital activity

Digital activity is understood as a human activity developed within a *digital economy*. The latter is characterized by profound changes in the perspective of economic development, made possible by the new technologies used on a large scale. Four main trends can be identified (Valenduc 2017):

– the strategic use of large amounts of data (otherwise known as "Big Data") produced by powerful algorithms;

– the industrial use of communicating and partly autonomous objects and machines (Industry 4.0) for the mass production of small customized series;

– the deployment of the notion of the network, as an organizing principle of life in society beyond the sphere of work, modifying the relationship with space and time;

– the use of online platforms replacing a new business model in the field of service delivery and the distribution of goods.

It is therefore, first of all, the business models of companies that are undergoing profound changes in a multitude of possible directions, taking traditional work organizations with them through a new and interconnected relationship with technology and shaking up whole sections of society as a central component of these business models. The sudden entrance of the smartphone onto the scene of our daily lives, for example, has put us in direct and permanent connection, via *a priori* harmless apps, with these business models. Whether it is buying a train ticket, reading a newspaper, listening to music, choosing a piece of clothing or ordering lunch, the slightest tap on an icon has become a central and now indispensable economic act for railway companies, media outlets, the music industry, goods distribution companies or home delivery platforms; it is important for them to motivate on one hand and respond on the other.

On the business side, the digital transition is characterized by new configurations accompanying these transformations, according to the principle, again an economic one, that technological innovations must be accompanied by organizational changes in order to generate productivity increases (Askenazy and Gianella 2000). In this context, the digital transition of work can be observed through "new regimes of subordination and new modes of work organization, [...] shift the lines of control and autonomy of subjects, [...] reformulate the conditions of commitment, responsibility, deliberation and the meaning of work" (Anact 2018, p. 2). In other words, the transformations of activity are not so much at play in the relationship that subjects maintain with digital tools (through their interfaces, for example) as in their relationships with the organizations that these digital developments contribute to in-depth restructuring. From this point of view, *digital* activity (literally: "that which relates to the fingers") bears its name unsuccessfully. While questions of the usability of digital tools and interfaces can and must be raised (particularly in terms of software ergonomics), digital activity cannot be reduced to this dimension alone. It is indeed the worker's relationship with his or her environment, that must be taken into account in the ergonomics of digital activity. And more broadly still, it is the contemporary relationship between humans and the world in which they live, produce and create that is at stake here and that a form of human ecology must be able to apprehend (Cazamian 2006).

However complex they may be, these changes in work must nevertheless be able to be anticipated as they are the result of concrete projects carried out within companies. Project leaders can thus consider them as a new perspective for those who work within these organizations, by inviting them to project themselves into their future digital activity (by simulating it) and to keep control of it (by allowing them to put it in their hands – not to say "on their fingers").

16.3. Simulating digital activity

For the subject acting in a situation, the work activity is always at the interface of a *prescription system* (made up of objectives to be achieved, rules and means, numerical or not), of *action logics*, pre-developed by the subject in the course of his or her past experience and of *events* bursting into reality, which he or she must face (Figure 16.1a). The activity developed on this occasion produces effects directed towards the accomplishment of the task, as well as towards a readjustment of the characteristics of the situation, within the scope of the margins of maneuver, which are exploitable by the subject (Van Belleghem 2018). These margins of maneuver are, however, generally limited by the rigidities of the prescriptive system (not least of which is the subordination link) and most often result in constrained adjustments to the subjects' action logic. Incidentally, these "constrained adjustments" may affect their health, safety and/or subjectivity.

Each of the elements of this situation can, in a project, be configured to form scenarios (prescription, action logics, events), prefiguring the conditions for carrying out the activity in the future that we wish to simulate (Figure 16.1b). These scenarios must be precisely posed. The future prescriptive system, in particular, must be able to account for the digital options it carries (which tools?, which technologies?, which purposes?, which interfaces?) as well as for the different technical and organizational choices they induce (which work organization?, which task distribution?, which information-sharing logics?, which roles should be given to each person in relation to these digital perspectives?). The event scenarios must describe the variabilities of reality that the subjects will be confronted with tomorrow and that they must be able to anticipate in the simulation. Action logic scenarios are generally made up, from the outset, of the repertoire of workers' individual and collective strategies developed in their past experience (ways of doing things, professional gestures, reasoning, professional regulations, stabilized social interaction norms, etc.), but will be expected to diversify and develop as the simulations progress (Falzon 2013).

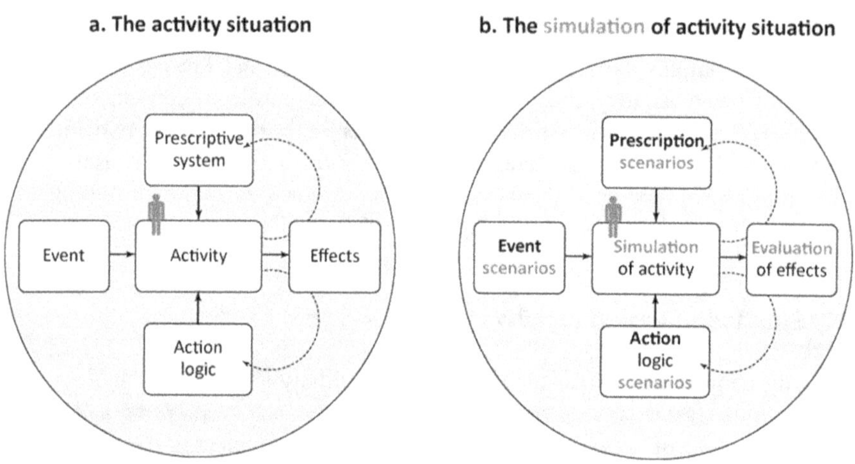

Figure 16.1. *The simulation situation, configurable mirror of the activity situation. For a color version of this figure, see www.iste.co.uk/bobillierchaumon/digital.zip*

For the elements being specified (even partially, at the start of the reflections, for example), the project supervisor must invite the workers to come and simulate their future activity in the prescriptive context being developed. This is an opportunity for them to come and test their action logic in light of the simulated situations, as well as to test the project in light of their action logic. A simulation device must be designed for this purpose, generally bringing together decision-makers, designers and workers around a simulation support (a model of the prescription system), allowing workers to "play" their activity to deal with specific event scenarios (Figure 16.2). It will also be useful, within the framework of social management of the intervention, to position the staff representative bodies in the monitoring of the approach.

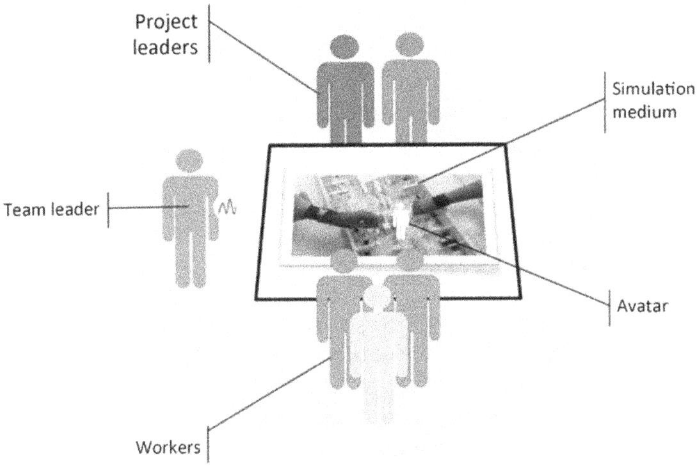

Figure 16.2. *The simulation system and its main actors. For a color version of this figure, see www.iste.co.uk/bobillierchaumon/digital.zip*

The simulation medium, in this context, must meet three requirements conditioning its design (Van Belleghem 2017):

– The first requirement (*Representation*) aims to represent the prescription elements in the form of scenarios that can be developed during simulations, to improve them. In predominantly spatial projects, for example, a model at a given scale of the spaces being designed, usually perfectly meets this representation requirement. In a predominantly organizational project, it will be a question of representing the system of rules in the course of the design. Even though these elements may seem abstract *a priori*, the challenge of representing them in a concrete form is crucial (e.g. by a model of the functional organization chart highlighting the nature of the links envisaged between its entities) because it must allow each person to make a representation shared with other actors of the project and to put it to the test of simulation. For the participant, this requires creativity and imagination to "demonstrate and manipulate" the structure and the beginnings of the organization under construction. This requirement proves to be just as fundamental for a predominantly digital project, where the representation of the "digital prescription system" will have to find its materialization.

– The second requirement (*Modification*) is intended to enable and facilitate collective modification of prescription scenarios. It is, above all, a question of enabling workers, in the same way as designers and decision-makers, to be able to act on the project by "physically" modifying the elements of the medium, in order to explore the scope of the possible prescription options and to propose alternatives to the options selected. Being able to move a partition on a building model or to bring one department closer to another on a functional organization chart offers power to act on the project, that must be allowed, or even sought. It is an essential condition to feed the project with the experience and intuition of employees, too rarely exploited despite its sometimes promising flashes of brilliance (Michel and Haeffner 2019). This requirement carries the seeds of employee support for the project and its intentions. Here, a digital project will find all the more interest when offering the participants the possibility of acting on the materiality of its scenarios, as its *a priori* virtual character could lead us to believe that it is elusive by nature.

– The third requirement (*Simulation*) enables a simulation of the activity. It expressly invites the speaker to have identified in advance, during the analysis of the work situations (in the diagnosis phase), the dimensions of the activity that are likely to evolve within the framework of the project. This identification should enable them to propose, during simulations, avatars adapted to the dimensions of the activity actually mobilized. In predominantly spatial projects, it is therefore not uncommon to use Lego© or Playmobil© characters manipulated on an architectural model, facilitating the embodiment of the subjects in the dimensions of the activity, specific to the logic of movement and use of space. The simulation of digital activities will similarly require a precise identification of the dimensions of the activity impacted, and the proposal of adapted avatars. Again, it should be noted that these dimensions can be varied and almost inevitably relate to very concrete work objects: technical systems, of course, as well as documents, files, written and/or oral information exchanges, customer interactions, etc. or even books, trays and carts (see below). The concrete nature of these work objects, capable of embodying the dimensions of the activity impacted by the digital project, makes them particularly apt to play the role of avatars to facilitate projection in future work. We must exploit this.

The two illustrations below, which deal with the support of projects in which the digital dimension was important (i.e. marking their belonging or their willingness to integrate into a *digital economy*), present significant examples of this.

16.4. Two illustrations of digital activity simulation

16.4.1. *A giant digital tablet to welcome customers*

The intervention, carried out for a large insurance company, concerns the support of a national project aiming at deploying a new concept of agencies within the commercial network. This concept was based on a whole series of technical, organizational and spatial evolutions organizing:

– the installation of a touch-screen digital panel in the reception hall inviting customers to register as soon as they entered the branch, enabling staff to be informed of their arrival (disappearance of the reception desk), to monitor and manage customer flows (instantaneous knowledge of the number of customers and the time since they arrived) and to organize the future availability of the various advisers;

– the provision of mobile computing tools (laptop, Bluetooth headset) allowing employees to connect where needed (processing of customer requests, administrative work) depending on the availability of space (individual reception areas, shared work areas), with the disappearance of dedicated offices;

– the application of new rules inviting all employees to welcome customers face-to-face and on the telephone as part of a consolidation of the local relationship, leading to a more anonymous use of the call center;

– the implementation of principles of dematerialization and the gradual disappearance of paper, in particular, by grouping the remaining printouts on a single printer;

– the use of an instant communication application between advisors;

– the merger of several agencies in certain territories.

This ambitious concept involved the client relationship, particularly in the changes it induced in the commercial and advisory act, as well as in the build up of coordination between the different players in the branch: client advisers, professional advisers, financial advisers and management. By implementing an organizational simulation system with the teams of several branches, the support aimed to facilitate the appropriation of the model while adjusting the technical and organizational principles envisaged. Analyses of work in the current configuration of the branches made it possible to identify the action logic at work among the employees: advance planning, real-time management of scheduled appointments and spontaneous visits, handling of telephone calls and e-mail requests, management of peaks of activity (particularly sudden rushes linked to local specificities – market day, proximity of the bus station, etc.), handling of administrative tasks, etc. Above all, the group's permanent mutual assistance could be highlighted, based on active vigilance by each of the advisers on what was happening in the branch and exchanges between visits, in order to schedule any readjustments that may have been necessary.

The simulation device designed explicitly aimed to "stage" the development of collective cooperation arrangements in the new agency concept (Figure 16.3). The choice of the simulation medium was based on a spatial model on a scale of 1:25 of the layout of the new agency proposed by the architects in response to the spatial concept. The digital display was particularly highlighted here. This medium was completed with personalized avatars representing the different consultants and clients. A programming grid for the day structured a planned sequence of events, while situation sheets, built from observations, allowed the user to initiate contexts or client requests that generated variability in the sequence of events. Staff members were invited to simulate their work by placing the avatars, representing them and processing the events and situations proposed to them.

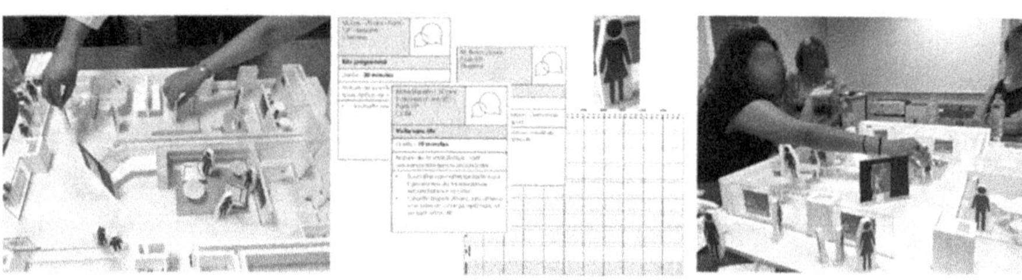

Figure 16.3. *Elements of the simulation device (model, situation sheets, programming grid for the day and avatars) and their handling. For a color version of this figure, see www.iste.co.uk/bobillierchaumon/digital.zip*

Immediately, the simulations engaged the employees to coordinate in order to manage the customers registering on the digital display. It was a matter of not only distributing specific requests according to the advisors present, but also managing the occupation of the different spaces over time. As the simulations progressed, new action logics emerged to facilitate the process of putting clients in touch with the correct advisor, to propose fair distribution principles, to ensure continuity of telephone calls or to structure new forms of mutual assistance through the use of a chat, including client appointments for help on sensitive issues.

The digital activity simulated here specifically took on new forms of cooperation between agency actors in a radically recomposed prescription system, but no less concrete than the one that preceded it. The rapid appropriation of the new agency concept by the staff, particularly in its digital dimension, was particularly facilitated by the tangible nature of the simulation system used.

16.4.2. *RFID chips for lending books*

Departmental media libraries are intended to make a collection of documents (books, CDs, DVDs) available to representative libraries throughout the region, which provide local loans to individual users in the department. This media library has recently modified its distribution logic, formerly carried out by a mobile library, by offering the librarian representatives the choice between a collection service on site or an online reservation followed by a shuttle delivery. After a few months of implementation, this organization was struggling to find its rhythm and meet the expected service challenges. The teams were experiencing difficulties linked to the increase in the number of loading areas, they warned of the inadequacy of the means at their disposal and pointed out unsuitable reception and service conditions to the librarian representatives.

The intervention requested by the departmental media library aimed to "rethink" its entire document delivery process to its partners as part of a participative and concerted approach with all employees. The use of new technologies, and more particularly RFID (radio identification tags for document identification), was mentioned as a possible avenue, but was still explored little.

Analyses of the work revealed that the principles of document distribution were not questioned by the staff, but confirmed that its implementation was cumbersome and restrictive, particularly in view of the number and diversity of handling operations it required (handling of books, boxes, unsuitable bags, etc.), as well as the associated IT operations (searching for documents, management of anomalies, etc.) and process constraints (redundant operations, loss of time, risk of mixing documents, etc.). The intervention was then oriented towards the redefinition of a document circulation process largely integrating RFID and nomadic technologies in order to reduce handling at all stages (both in the internal organization of the department and in partner libraries), to give consistency to the distribution logics and to refocus cooperation between the consulting librarians and the librarian representatives on to advice, rather than on the handling of books and other media.

The collective reflection mechanism was based on two tangible media. The first enabled the agents to redefine the content of each work step constituting the future document distribution process (Figure 16.4). It was a medium for the *scripting* of the expected prescription system.

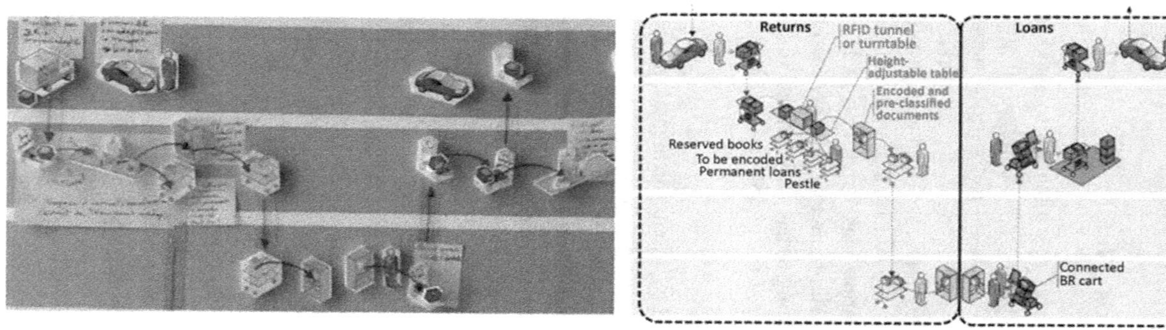

Figure 16.4. *The medium for the scripting of the future process and its formalization following the work (extract). For a color version of this figure, see www.iste.co.uk/bobillierchaumon/digital.zip*

The second medium schematically represented all the spaces crossed by the document distribution circuit, from the media library shops to the shelves of the representative libraries, including the reception and access area to the collections, the car parks (where the loading and unloading of book bins was carried out) and even the departmental roads connecting the various buildings to each other (Figure 16.5). This was the *simulation* medium on which agents were invited to "play out" their future activity, by manipulating different avatars representing their "work objects": miniature books placed in miniature bins and themselves positioned on miniature carts or in miniature vehicles, depending on the stage of the process. This representation made it possible to evaluate the value of the choices made in the process with regards to the physical handling of book bins and the associated loading areas.

Figure 16.5. *The simulation medium, made up of "work object" avatars (books and "RFID chip" bins, connected carts, etc.). For a color version of this figure, see www.iste.co.uk/bobillierchaumon/digital.zip*

The prospect of using the RFID technology here offers new perspectives that the media system (scenario and simulation) made it possible to explore widely and systematically. The aim of the working group was to limit the handling of books as much as possible. This objective was met by proposing to equip them with RFID tags and to equip the group of representative libraries with transport bins. For the "Loan" process, a "connected" cart was made available to the librarian representatives, enabling them to carry out the computerized loan as soon as the book was taken from the shelf and positioned in the bin, thanks to an automatic RFID tag recognition system. For the "Return" process, the containers provided avoided bringing back borrowed documents in boxes or bags that were not suitable for transporting books and facilitated the registration of documents by engaging them in a tunnel that recognized all the RFID tags in a single process. The bins themselves were managed like books, that is, equipped with RFID chips and loaned to the representative libraries that had to return them. This principle eliminated all the intermediate loading/unloading of the bins, helped to ensure the smooth circulation of documents in a single container and reduced the risk of mixing documents. The simulations made it possible to envisage the support of the librarian representatives during the Return-Loan process by the consulting librarians who, freed from many of the old tasks of handling books (placing them in the bins, scanning on the table, management of the representative librarians' boxes and bags, etc.), could concentrate on their advisory role.

The digital activity here took on a very singular form: effectively part of a digital economy in which the management of the exchange of documentary collections between the departmental media library and the representative libraries was, above all, a question of feeding computerized databases that were networked. It was in the physical handling of books that the reliability of the information was at stake, as well as the service provided to the user (purpose of the activity). It was not so much the materiality put at the service of digital technology, as the digital technology put at the service of the "book" object, the latter having been able to defend an original paper format within a media offer that is now being massively swept away by the wave of streaming.

16.5. Conclusion

We have to admit: digital activity is not connected to digital technology. At least not in the sense in which it is generally understood, that is, virtual, abstract, even magical. On the contrary, like any activity, it is linked to reality in its most tangible form. It is in this material relationship that it is able to manipulate it, to direct it and, *in fine*, to keep control of it. Consequently, by helping workers to plan for their digital activity, particularly in the context of reorganization projects, the fears which we know they can arouse, invites them to cling to this materiality. In particular, it is a matter of demystifying a whole set of representations of digital technologies and their supposedly elusive dimensions.

As a configurable mirror of the business situation, the simulation situation must reflect the tangible nature of future work, even in an advanced digital economy. Numerical simulation media can be envisaged in this objective. The lockdown period during the health crisis of 2020 has also provided us with the opportunity to develop "remote simulation" devices based precisely on the use of interactive tools that allow participants gathered in videoconferencing to play out their activities remotely. The development of these tools has focused on the best possible representation of the tangible elements of work performance (workspaces, situation and event maps, temporal freezing of the work day, significant avatars, work objects and even masks and bottles of hydroalcoholic gel to anticipate their generalized use in future work situations). However, there is nothing to indicate, as soon as the possibilities of grouping participants face-to-face are combined, that these technologies are always the most capable of meeting the methodological requirements mentioned above (being able to represent, being able to modify, being able to simulate). If this were not the case, the participant would have a responsibility: to know how to resist technological esthetics in order to design devices capable of constructing the experience of the materiality of the work. Paper, cardboard and Playmobil© characters sometimes do more, in this sense, than 3D representation and virtual reality glasses. Perhaps also because reality is, precisely, never virtual.

16.6. References

Anact (2018). La transition numérique, promesses et menaces pour "l'expérience travailleur". *Explo Qualité de vie au Travail*, (1) [Online]. Available at: https://www.anact.fr/la-transition-numerique-promesses-et-menaces-pour-lexperience-travailleur.

Askénazy, P. and Gianella, C. (2000). Le paradoxe de la productivité : les changements organisationnels, facteur complémentaire à l'informatisation. *Economie et statistique*, (339–340), 219–242.

Cazamian, P. (2006) De l'approche globale en ergonomie et en écologie humain. In *A quoi sert l'ergonomie ?*, Aletcheredji T. and Heddad N. (eds). Octarès, Toulouse, France.

Falzon, P. (ed.). (2013). *Ergonomie constructive*. Presses Universitaires de France, Paris, France.

Michel, J.-C. and Haeffner, I. (2019). La simulation du travail comme soutien à l'arbitrage dans une démarche de co-construction. *Speech, 54th SELF Conference*, Tours, France.

Valenduc, G. (2017). Au doigt et à l'œil. Les conditions de travail dans des environnements digitalisés. *HesaMag*, (16), 12–16.

Valenduc, G. and Vendramin, P. (2016). *Le travail dans l'économie digitale : continuités et ruptures*. ETUI, Brussels [Online]. Available at: http://hdl.handle.net/2078.1/171855.

Van Belleghem, L. (2017). What are the design requirements for an organisational simulation support? *Proceedings of ACE ODAM 2017*, Banff, Canada [Online]. Available at: https://bit.ly/2JQVxf9.

Van Belleghem, L. (2018). La simulation de l'activité en conception ergonomique : acquis et perspectives. *Activité*, 15(1) [Online]. Available at: http://journals.openedition.org/activites/3129.

17

Managing Technological Change

At the societal level, as in organizations, technological change is ambivalent. While it is accepted that it is a decisive factor in economic growth, it is no less true that it can also be an amplifier, or even a catalyst, of inequalities. This chapter begins by outlining some of the threats and opportunities related to the digitalization of organizations, before showing how social and responsible support can help to remedy the main threats. It then goes on to present a change management model that takes into account the specificities of technological change.

17.1. Introduction

The preceding chapters contain – directly or indirectly – either through case studies or models with a prescriptive dimension, elements of an answer to the question "How can we drive technological change?". For this purpose, we have chosen a particular summative angle: that of strategic change (Hafsi and Fabbi 1997). The guiding idea of our analysis is that the introduction of a new technology at organization level is a complex change that can challenge internal balances. It affects the business, the employees and the organization as a whole. As such, it is a source of both threats and opportunities. We will focus on these notions.

We begin by presenting some of the threats and opportunities related to the digitalization of organizations, before presenting how social and responsible management of these new technologies can address the main threats. Finally, we present a change management model that takes into account the specificities of technological change[1].

Chapter written by Clotilde CORON and Patrick GILBERT.

1 Developments on the reflections outlined here can be found in Coron and Gilbert (2019).

17.2. Digital transformations, sources of threats and opportunities

Technological change has always been seen in an ambivalent way. It is sometimes seen as a threat and sometimes as an opportunity. The legions of technophiles and other "geeks" have always been opposed by the no less numerous techno-critics, from the rejection of machines to the contestation of ICTs. Digital transformations are re-actualizing these controversies. Table 17.1 provides some examples. It is interesting to note that for each threat, an opportunity can be highlighted (and vice versa). Rather than following one side or the other, we take them both seriously, observing the double play of technological change, between light and shadow.

These reflections can be related to Georges Friedmann's questioning of the power of technology, which he spoke of in his work on industrial machinery (1946). In analyzing the damage caused by Taylorism, he showed that if technical systems are not neutral, they also have extreme flexibility.

Domains	Threats vs. opportunities	Consequence highlighted in the literature	Transposition to digital transformation	Examples
Humanity	Threat	Technical ideology (Ellul 1980)	Discourses exaggerating the merits of digitalization, without taking into account its negative aspects	Implementation of a digital technology whose qualities have been exaggerated and whose difficulties have been masked
	Opportunity	Social progress (Sauvy 1980)	New technologies responding to social needs	Cloud-based service for the homeless
Social Justice	Threat	Inequalities and exclusion	Digital divide (DiMaggio and Hargittai 2001; van Dijk and Hacker 2003) Algorithm discrimination (O'Neil 2016)	Implementation of digital tools without adequate staff training Use of a recruitment algorithm (e.g. automation of CV pre-selection) that reinforces existing human biases

	Opportunity	More inclusive society	Digitalization offers new working tools for the inclusion of people with disabilities, for example Thoumie (2004)	Digital tools to summarize text vocally (for visually impaired people, for example)
Volume of jobs	Threat	Destruction of jobs (Cépède *et al.* 1951; Simon, 1960)	Automation of certain tasks due to digitalization (Frey and Osborne 2017)	Automation of part of the profession of notary clerks under the effect of digitalization
	Opportunity	Job creation by shifting labor from one sector to others (Sauvy 1980; Schumpeter 1911: creative destruction)	Emergence of new jobs and tasks	Professions related to community management and the Internet in general
Job content	Threat	Prescriptive technique, alienation of the human being (Weil 2002)	Management by algorithms (Lee *et al.* 2015; Lee, 2018)	Uber drivers or Deliveroo couriers whose work is entirely prescribed by algorithms
	Opportunity	Technique to help with the realization and work	Digital techniques that provide assistance to humans (e.g. using Big Data as a decision aid; see Mayer-Schönberger and Cukier 2013)	Decision support software, modeling software
Working conditions	Threat	Stress at work	Hyper-connection, porosity of personal and professional lives (Bawden and Robinson 2009)	Burnout related to the intensive use of new technologies
	Opportunity	Improvement of working conditions	Improvement of working conditions and social climate	Digital tools to facilitate telework and telecommuting (Vayre 2019)

Table 17.1. *Ambivalence of digital technologies*

Firstly research has highlighted the dangers of the "technicist ideology", which presents technology only in a positive light and sees it as a necessity. Ellul (1980) is one of the representatives of this trend. The technicist ideology is just as present concerning digital technology, often seen as a source of progress and not always questioned. For example, companies are launching into setting up corporate social networks on a mass scale, sometimes without questioning the value of this type of tool. Conversely, the technique can also meet a need and thus ensure a form of social progress (Sauvy 1980). This is particularly the case when new technologies provide solutions to the needs of people in precarious situations, who are usually left out of technological progress.

The technique is also presented as a source of exclusion, particularly when its use requires skills unequally distributed among the population. With regard to the digital divide, van Dijk and Hacker (2003) show that, although the ownership differentials of digital tools (computers, smartphones, etc.) tend to decrease, the use and skills differentials remain significant. This differential depends in particular on age, income, level of education and gender. DiMaggio and Hargittai (2001) point out how the digital divide covers several types of inequalities: inequalities in equipment, in autonomy of use, in skills, in social support and in the objective that the use of digital technology aims to achieve. Technology can also be a source of exclusion when it leads to discrimination. O'Neil's work (2016) on Big Data shows how digital technology can indeed reproduce, as well as aggravate, already existing forms of discrimination and exclusion. Conversely, new technologies can also be used to promote inclusion. To give just one example, technical progress in medicine, in particular, has led to considerable progress in the inclusion of people with disabilities. Digital technologies are not to be outdone (Thoumie 2004). For example, some software now makes it possible to summarize written text vocally, which is a considerable help for visually impaired people.

Although economists such as Simon (1960) believe that technological progress is compatible with full employment, Cépède *et al.* (1951) stressed the effect of technology on employment very early on, taking the example of farm workers who saw their jobs disappear as a result of the mechanization of agriculture. Frey and Osborne (2017) have transposed this observation to digitalization, focusing on the occupations most threatened by digitally related automation. The probability of automation of a professional task thus depends on three factors: the degree of social intelligence, the degree of creativity and the degree of precision in the handling it requires. These three factors are can be broken down into sub-factors:

– social intelligence: perception of others and understanding of their behavior, negotiation, persuasion, assistance to others;

– creativity: originality, artistic mastery;

– precision: dexterity of fingers, dexterity of hands, cramped work spaces.

They thus demonstrate that the work of notary clerks is under serious threat from the automation associated with digitization. Indeed, this relatively routine and administrative work requires little social intelligence, creativity and manual dexterity. Conversely, new technologies can also become a source of job creation, particularly in the digital field. Numerous professions have thus emerged in recent years with in this field, such as those related to content management on the Internet, the analysis of data resulting from digitization or the management of online communities. This phenomenon of spilling over of the workforce from one sector to another, in connection with technological change, is not new (Sauvy 1980).

Researchers, philosophers and sociologists in particular, have pointed out that technology can have an alienating effect when it prescribes entirely the tasks and activities of workers: production lines, for example (Weil 2002). In the digital field, the notion of "algorithmic management" has recently been introduced to account for situations where tasks are entirely prescribed by algorithms (Lee *et al.* 2015; Lee 2018). This is the case, for example, of Uber drivers or Deliveroo couriers, who receive their work instructions on an application on their smartphone and can therefore perform their work entirely without any human manager or supervisor. However, in some cases, technology can also make work easier, while leaving the individual room for maneuver. This is the case, for example, with decision support software, which today incorporates artificial intelligence and makes it possible to summarize a large body of information in order to assist human decision-making. Mayer-Schönberger and Cukier (2014) give many examples of how Big Data can be mobilized in the service of, for example, medical diagnosis or maintenance diagnostics

Many researchers, particularly psychologists and sociologists, have studied stress in the workplace. Digital technology can contribute to this stress when it is used intensively and leads to a loss of boundaries between private and professional life: it becomes possible to be permanently reachable for professional reasons, even outside of working hours and during holidays. Bawden and Robinson (2009) thus refer to information overload, the anxiety linked to permanent and easy access to information, for example. In the most pathological cases, this can lead to burnout (Bobillier Chaumon 2017). Conversely, digital technologies are also sometimes presented as improving working conditions, by making certain tasks less arduous or by promoting remote working and telework (Vayre 2019).

17.3. Social and responsible management of new technologies

This presentation of threats and opportunities highlights the importance of managing new digital technologies responsibly. The concept of responsible innovation sheds light on what responsible management of digital technologies could be.

17.3.1. *Social and responsible innovations*

Social and responsible innovation is a response to poorly or inadequately met social needs, taking into account its negative externalities and seeking to limit its effects (Laurent *et al.* 2018). This response generally combines a technical dimension and a social or societal dimension (see example in Box 17.1).

> In the field of recycling, cell phone reconditioning and resale workshops employing people who have long been out of work (e.g. long-term unemployed) are increasingly developing. This innovation responds to a collective need, that of recycling devices whose production proves to be harmful to the environment, through a technical (e.g. component repair) and social (e.g. hiring people out of work) solution. In a different field, the creation and production of insulated shelters for homeless people have both a technical and a social dimension, and address the collective problem of poverty and poor housing.

Box 17.1. *Social and responsible innovation, a combination of technical and social elements*

Social innovations can therefore be described as responsible, insofar as they take into account all the effects they may have on society and individuals, and they seek to limit negative effects.

17.3.2. *Digital and responsible technological innovation*

Here we propose the concept of responsible technological innovation, which refers to technological innovation that incorporates the characteristics of social and responsible innovations, such as meeting a need and combining a technical and a social dimension. Responsible technological innovation makes it possible to seize the opportunities presented in the first section, while also limiting the threats.

Thus, responsible technological innovation is characterized, like responsible social innovation, by the fact that it responds to a collective and social need. This limits the power of technicist ideology and situations where technology is mobilized without real necessity. Moreover, it is particularly active in areas such as the fight against exclusion, poverty, unemployment and crime, which are all threats associated with new technologies. It can also be used to improve working conditions and social progress, and in this way, it makes it possible to seize the opportunities listed in Table 17.1.

Responsible technological innovation lends itself particularly well to the field of new and digital technologies. Table 17.2 therefore gives examples of digital innovations that limit threats and seize opportunities.

Domains	Threats	Opportunities	Example of responsible technological innovation
Humanity	Technical ideology	Social progress	Smartphone applications to promote the resale of unsold food products
Social justice	Inequalities and exclusion	More inclusive society	Applications on smartphones that aim to compensate for certain disabilities (e.g. vocal summary of texts, textual summary of speech, etc.)
Volume of jobs	Destruction of jobs	Job creation	Recycling structures for digital tools (computers, smartphones) employing people who have long been out of work
Job content	Prescriptive technology, alienation of the human being	Technology to help with the realization and work	Applications to facilitate administrative procedures (e.g. applying for leave)
Working conditions	Stress at work	Improvement of working conditions	Tools facilitating remote work as well as the right to disconnect

Table 17.2. *Responsible technological innovation in the digital domain*

Thus, the first threat refers to technical ideology, and the corresponding opportunity refers to social progress. Innovations such as smartphone applications that aim to promote the resale of soon-to-expire food products to preferential products meet two social needs: limiting food waste and providing access to food products at low prices.

With regard to discrimination and exclusion, the integration of people with disabilities in the workplace can be facilitated by digital technologies that aim to compensate for certain disabilities. For example, automatic text-to-speech techniques can compensate for a visual deficit, and automatic written transcription of sound extracts can compensate for a hearing disorder.

New technologies can, as we have seen, help to destroy as well as create jobs. In this field, responsible technological innovations can aim to create jobs for people in difficulty, for example, in the recycling of electronic objects.

They can also be both alienating, when they prescribe work, and liberating, when they help the worker and give him or her room for maneuver and freedom. For example, some computer-based or even smartphone applications are designed to simplify administrative processes and procedures: filing applications for leave, expense claims, changing addresses or bank details, reporting sick leave, etc.

Finally, they can be a source of stress or, on the contrary, a source of well-being at work. For example, tools that facilitate remote working are a double-edged sword: they can provide a certain degree of comfort by making it possible to avoid commuting to work, for example, but they can force individuals to be permanently connected to their work devices. Therefore, applications that grant the right to disconnect can be useful: the possibility of working in "offline" mode, the possibility of indicating that we do not wish to be disturbed, the possibility of cutting the professional line on a cell phone with a dual SIM card, for example.

17.4. A model for responsible leadership of technological change within organizations

How, then, can technological change be conducted in a way that respects the principles of responsible technological innovation? The literature on change management is particularly abundant. There is a persistent practice of distinguishing between the driving forces of change, and, in particular, of distinguishing between technological change and organizational change. To be frank, this tradition is outdated, because all technological change obviously has an organizational dimension, and technological change does not go without organizational change management.

17.4.1. *Managing organizational and technological change*

Many authors have examined change management, either by studying the practices of organizations in this area or by proposing practices that are adapted and presented as more effective than others, with a normative aim.

Thus, Philips (1983) proposes a four-phase change management model, which he presents as critical success factors: creating a consensus on the need for change; defining the target timeframe; defining the concrete actions to be implemented to achieve this objective; and stabilizing and institutionalizing the new organization. This prescriptive model remains broad enough to allow each organization to adapt the content of each step to its context. This is all the more important since other authors, such as Pettigrew (1987) or Weick and Quinn (1999), stress that, on the contrary, the management of change is highly contingent. It depends in particular on the following factors: the extent of the change sought, organizational characteristics such as age, size, organizational model and cultural inertia.

Other authors have been specifically interested in the notion of resistance to change, whether this resistance comes from organizational power struggles (Crozier and Friedberg 1992) or from more individual psychological factors (Coch and French 1948), for example.

In the more specific area of technological change, authors have examined the factors structuring user acceptance of a new tool or technology. For example, Davis (1993) proposed a technology acceptance model (TAM), in which perceived usefulness and ease of use are the main determinants of technology acceptance. This model has subsequently been amended by a series of other works (Bobillier Chaumon and Dubois 2009); see Chapter 16 on the conditions of acceptance of technologies in and through activity.

17.4.2. *Responsible digital change management*

Finally, a model for driving technological, and more specifically, digital change could be largely based on conventional change management models. However, such a model would not necessarily integrate the "responsible" dimension, which seems key for digital technologies, in view of the threats and opportunities they represent. Table 17.3 summarizes the opportunities and threats in Table 17.1, from which we deduce the implications for digital change management that would aim to seize these opportunities while limiting the threats.

For example, in order to avoid the pitfalls of technical ideology while promoting social progress, this change management could promote technologies that meet a real need, following the example of social and responsible innovations. Training all workers in the new tools could reduce the risks of a digital divide, and digital tools such as algorithms could be used to reduce discrimination if this objective is included in the design of these tools from the outset. In order to reduce the harmful effects of job losses, it might be necessary to limit, as much as possible, the number of job disruptions, given that digital technology is also supposed to create jobs, and to support the employees concerned. Promoting digital tools that leave individuals room for maneuver would also make it possible to limit the risk of alienation, while, at the same time, increasing individual freedom. In terms of working conditions, supporting workers in the use of tools, authorizing the right to disconnect and encouraging teleworking seem necessary.

Domains	Threats	Opportunities	Implications for change management
Humanity	Technical ideology	Social progress	Ensure that the tools proposed meet a real need
Social justice	Inequalities and exclusion	More inclusive society	Systematically train all staff in the new tools Take diversity issues into account when designing algorithms
Volume of jobs	Destruction of jobs	Job creation	Limit job terminations, support the employees concerned
Job content	Prescriptive technology, alienation of the human being	Technology to help with the realization and work	Promote digital tools that leave individuals room for maneuver
Working conditions	Stress at work	Improvement of working conditions	Supporting employees in the use of digital techniques to avoid burnout, to authorize the right to disconnect, to encourage teleworking, etc.

Table 17.3. *Limiting threats and seizing opportunities*

We then consider the change management that would respect the principles of this table as "responsible digital change management". Indeed, it would be change management that would incorporate the principles of social and responsible innovation presented above.

17.5. Conclusion

Finally, new technologies offer opportunities that should not be obscured by the threats they present. To seize these opportunities while limiting these threats, we propose the concepts of responsible technological innovation and responsible digital change management. It is, in fact, a question of applying the principles of social and responsible innovation to the digital domain.

This involves, in particular:

– promoting digital tools that leave individuals room for maneuver, which aid, rather than prescribe, work;

– using digital tools for a more inclusive society;

– giving priority to tools that improve the working conditions of individuals, without presenting additional sources of stress;

– assessing the merits of digital change by studying its effects on individuals and not hesitating to if necessary.

These principles are compatible with broader theoretical approaches, such as those developed by the psychology of organizational change (Oreg *et al.* 2013) or the sociology of change (Bernoux 2004), as well as with intervention methodologies developed by occupational psychologists, to support the deployment of new technologies (Pichot *et al.* 2018).

17.6. References

Bawden, D. and Robinson, L. (2009). The dark side of information: Overload, anxiety and other paradoxes and pathologies. *Journal of Information Science*, 35(2), 180–191.

Bernoux, P. (2004). *Sociologie du changement*. Le Seuil, Paris.

Bobillier Chaumon, M.E. (2017). Du rôle des TIC dans la transformation digitale de l'activité et de la santé au travail. *La revue des conditions de travail,* 6, 16–24.

Bobillier Chaumon, M.E. and Dubois, M. (2009). L'adoption des technologies en situation professionnelle : quelle articulation possible entre acceptabilité et acceptation ? *Travail humain*, 72(4), 355–382.

Cépède, M., Picard, F., Tugault, M.J., Beaulieu, D., and Carrel, J. (1951). Section II : les conséquences de la motorisation. *Économie rurale*, 3(1), 30–43.

Coch, L., and French Jr., J.R.P. (1948). Overcoming resistance to change. *Human Relations*, 1(4), 512–532.

Coron, C. and Gilbert, P. (2019). *Technological Change*. ISTE Ltd, London, and Wiley, New York.

Crozier, M.F. and Friedberg, E. (1992). *The Actor and the System: The Constraints of Collective Action*. SAGE Publications, New York.

Davis, F.D. (1993). User acceptance of information technology: System characteristics, user perceptions and behavioral impacts. *International Journal of Man-machine Studies*, 38, 475–487.

van Dijk, J, and Hacker, H. (2003). The digital divide as a complex and dynamic phenomenon. *The Information Society*, 19(4), 315–326.

DiMaggio, P. and Hargittai, E. (2001). From the "Digital Divide" to "Digital Inequality": Studying internet use as penetration increases. Working Paper, Princeton University, New Jersey.

Ellul, J. (1980). *The Technological System*. Continuum, New York.

Frey, C.B.F. and Osborne, M. (2017). The future of employment. *Technological Forecasting and Social Change*, 114, 254–280.

Friedmann, G. (1946). *Problèmes humains du machinisme industriel*. Gallimard, Paris.

Hafsi, T. and Fabi, B. (1997). *Les fondements du changement stratégique*. Éditions Transcontinental, Montreal.

Laurent, B., Baker, M., Beaudouin, V., and Raulet-Croset, N. (2018). *Innovation et participation. Approches critiques*. Presses des Mines, Paris.

Lee, M.K. (2018). Understanding perception of algorithmic decision: Fairness, trust, and emotion in response to algorithmic management. *Big Data & Society*, 5(1), 1–16.

Lee, M.K., Kusbit, D., Metsky, E., and Dabbish, L. (2015). Working with machines: The impact of algorithmic and data-driven management on human workers. *Proceedings of the 33rd Annual ACM Conference on Human Factors in Computing Systems*, Seoul, South Korea.

Mayer-Schönberger, V. and Cukier, K. (2013). *Big Data: A Revolution That Will Transform How We Live, Work, and Think*. Houghton Mifflin Harcourt, New York.

O'Neil, C. (2016). *Weapons of Math Destruction: How Big Data Increases Inequality and Threatens Democracy*. Crown, New York.

Oreg, S., Michel, A., and By, R.T. (eds) (2013). *The Psychology of Organizational Change: Viewing Change from the Employee's Perspective*. Cambridge University Press, Cambridge.

Pettigrew, A.M. (1987). Context and action in the transformation of the firm. *Journal of Management Studies*, 24(6), 649–670.

Phillips, J.R. (1983). Enhancing the effectiveness of organizational change management. *Human Resource Management*, 22(1–2), 183–199.

Pichot, N., Quiguer, S., and Somat, A. (2018). Un cadre psychosocial d'intervention pour accompagner le développement et le déploiement d'une technologie nouvelle. *Psychologie du travail et des organisations*, 24(4), 355–373.

Sauvy, A. (1980). *La machine et le chômage : le progrès technique et l'emploi*. Dunod/Bordas, Paris.

Simon, H. (1960). *The New Science of Management Decision*. Prentice-Hall, Englewood Cliffs, New Jersey.

Thoumie, P. (2004). Recherche technologique et diffusion de l'innovation au service du handicap. Report, French Ministry of Youth, Education and Research.

Vayre, E. (2019). Les incidences du télétravail sur le travailleur dans les domaines professionnel, familial et social. *Le travail humain*, 82(1), 1–39.

Weick, K.E. and Quinn, R.E. (1999). Organization change and development. *Annual Review of Psychology*, 50, 361–386.

18

Exploring the Situated Acceptance of Emerging Technologies in and Concerning Activity: Approaches and Processes

What is the basis for the acceptability of technologies? What are the different perspectives for analyzing the acceptability of technical devices? What are the limits and contributions of these models? When the technologies do not yet exist or are in the re-design phase, how can we explore the adoption of these technologies which should contribute to an acceptable future for the transforming activity? The aim of this last chapter is to understand the different approaches to technological acceptability (practical, social and situated) and to set out, from the perspective of the clinical element of use, the approaches that can be used to assess the impact and acceptable conditions of the implementation of new technologies.

18.1. Introduction

Assessing the acceptability of technologies is undoubtedly one of the most important issues in the design and deployment of technologies, since the objective is to ensure that these tools will be used by users, according to their intended use. It is important for those in charge of the development and selection of technologists to identify the factors and conditions that may lead future users to adopt or reject the devices that will be implemented; whether for economic, socio-organizational or psychosocial reasons:

Chapter written by Marc-Eric BOBILLIER CHAUMON.

1) economic reasons first, because it is necessary to be certain that the return on investment is profitable for the company, that is, that users actually use the device for the tasks and the intended uses, taking into account the costs of development, purchase and integration (Brow *et al.* 2002): this can range from several thousand euros, or even millions of euros for integrated management or artificial intelligence software packages;

2) then socio-organizational factors, because the implementation of these technologies will lead to major reconfigurations of the *work to be done* (its nature, its aims, the quality criteria, etc.) and *the ways of doing it* (the methods of exercise as well as the principles of collaboration and management) (Vuori *et al.* 2019). It is therefore necessary to ensure that these various transformations remain bearable for the work system as a whole and that they have beneficial effects for the performance of the company and the efficiency of employees;

3) this ultimately leads to an assessment of the psychosocial impacts that these new working arrangements may generate, in particular, on the well-being of employees and their quality of life at work. Indeed, these technological devices contribute to the conditions in which work is carried out. They can therefore promote or, on the contrary, prevent work, and thus affect the quality of work, which is known to be a major condition for employee recognition and well-being (Clot 2008). The individual may therefore find himself or herself weakened by unsuitable technologies that will require additional cognitive and/or subjective commitment, in order to perform his or her work. They may also feel mistreated or useless when the device appropriates the core of the job – the one that gives meaning to the work – or when the technology requires actions that are contrary to the quality criteria and rules of the job that are theirs (Bobillier Chaumon 2017).

Assessing the acceptability of technologies by employees is therefore a means of protecting against possible abuse, misuse or rejection of technologies and of encouraging their appropriation by users. However, as we shall see in this chapter, it is not a matter of getting subjects to accept the technologies *at all costs, whatever the cost*, but rather of making them acceptable, in other words, of creating acceptable conditions and favorable resources for their adoption. This is what we have called situated acceptance, which is part of a more general approach that is the responsibility of the clinic of use. First, we will explain the different perspectives relating to the acceptability of technologies (practical, social and situated acceptance), specifying the modalities of situated acceptance, which in our opinion, are more capable of restoring the favorable or unfavorable effects of the use of technologies on different dimensions of the system of professional activity. We will then discuss the methods that can be used to evaluate this situated acceptance.

18.2. Models of technological acceptability: outlines and approaches

Let us start by defining what is meant by the generic concept of technological acceptability. In a process of design, innovation, deployment and/or technological change, it is about predicting, measuring, explaining and improving user acceptance of technologies. This is done by identifying and controlling the factors, conditions and/or motivations that may lead individuals to want to use or not use new technological devices.

In this perceptive, three main approaches can be distinguished, depending on the state of maturation of the innovation and the level of development and diffusion of the tool in the activity.

18.2.1. *Practical or instrumental acceptability*

This first type of acceptability aims to design ergonomic innovations that are as useful, usable and accessible as possible, that is, compatible with the needs and specificities of individuals and the characteristics of their activities. We thus seek "to design not only an adapted product but also a relationship adapted to the product" and finally an acceptable "user experience" for the user (Barcenilla and Bastien 2009, p 2). Therefore, to be acceptable, a technology must "strive to amplify, by making them more fluid, the actors' capacities to make relevant information available at the right time, to promote the capitalization of knowledge by enhancing the value of information, and to limit cognitive and emotional overload" (Neveu *et al.* 2012, p 79). This practical acceptability, as defined by Neilson in 1993 in his "*systems acceptability*" model, is investigated by various ergonomic orientations (cognitive, software and design; Alexandre *et al.* 2018) or by ergonomic psychology, particularly through the symbiosis model (Brangier *et al.* 2009). The measurement of this practical acceptability is based on the ergonomic qualities of the device and on the evaluation of the user experience (Brangier and Bastien 2010; Hassenzahl *et al.* 2013). It provides concrete solutions for the design and evaluation of systems (ergonomic criteria, heuristics, user tests, etc.) (Leduc *et al.* 2014).

It may also be added that practical acceptability tends to increasingly address "hedonic" dimensions from heterogeneous conceptual approaches: the concept of fun or "funology", esthetics, pleasure, emotion, hedonic value or quality (Norman 2004; Dupré *et al.* 2018). In these approaches, using a product means satisfying a certain need (utilitarian dimension) to achieve a certain pleasure (hedonic dimension). The perception of the instrumental qualities of the innovation (perceived usefulness and usability) and non-instrumental qualities (esthetics, value conveyed) then induce decisive emotional reactions in the acceptance and appropriation of the innovation (Monéger *et al.* 2018).

While practical acceptability is a necessary precondition for the use of technologies, this approach can sometimes be perceived, as Barcenilla and Bastien (2009) indicate, as a "toolbox" approach, by its ambition to systematically identify available resources and methodologies to the detriment of a more analytical and explanatory approach to the work and usage situations.

18.2.2. *The social acceptability model*

The social acceptability model is undoubtedly the most used and mobilized approach in prospective studies concerning technology adoption processes, with nearly 2000 citations referenced in scientific articles in 2011 alone, according to Hsiao and Yang (2011). Schematically, the user is perceived as a "rational actor", who applies a probabilistic reasoning on the future of his or her use by evaluating the systemic benefits and risks of the use of the device, without any confrontation with the reality of the activity.

Specifically, social acceptability is the initial step in the technology adoption process and expresses the potential degree of acceptance of a technology (being designed or installed) by users. Attention is then focused on the personal representation of the technology by its potential users. It refers to the anticipated perception of a potential risk or benefit, in the short or long term – related to values, opinions, needs, attitudes, social norms, etc. – that may or may not be favorable to the technology developed (prognosis of use).

Many theoretical models have sought to account for the determinants of technological acceptability, such as the technology acceptance model (TAM) developed by Davis (1989) and its various variants (TAM 2, 3, UTAUT 1 and 2) (Atarodi *et al.* 2019). Two recurring factors seek to predict usage intent: perceived utility and ease of use. Other variables have subsequently been introduced, such as user characteristics (age, gender, professional category, years of experience, etc.), organization (support from managers, communication policy, social influence, etc.) and the characteristics of the device itself (proposed functionalities, task/technology suitability, etc.) (Bobillier Chaumon and Dubois 2009; Dubois *et al.* 2015).

These models seek to predict future user behavior based on a certain number of variables, in order to anticipate the reasons for the rejection or adoption of a technology. They emphasize the crucial and decisive role of social cognitions and consider that technology adoption is a rational computational process. The idea is that positive perceptions will help create a favorable context for system ownership, while negative perceptions lead to failure. Measures of acceptability therefore have a dual purpose: to improve the attributes that reduce the likelihood of rejection and to increase the likelihood of the attributes that lead to adoption of the system.

The methodologies used to assess these perceptions are almost exclusively quantitative and normative measures based on questionnaires and measurement scales. However, the pitfall of these evaluation techniques is oversimplification. This leads to considering the planned or expected uses as an actual experience, whose overall level of impact could be predicted and evaluated, as we would do for simple perceptual or cognitive stimuli, without taking into account the general conditions which the conduct is carried out in.

This rationalist prism puts representation at the origin of any possible action, in terms of adoption or rejection of technologies. This gives the illusion of an autonomous power of social cognition, which can be measured, modeled and, above all, controlled through mathematical models; by acting on this or the individual variable through behavioral incentives or corrections. This positivist perspective therefore leads us to favor the analysis of acceptability solely on the basis of the subjects' perception of it, and not to understand what they actually do with these technologies. In other words, there would be, on one hand, a normative conception of – social – acceptability which, in the name of a certain "scientific" objectivity, seeks to "represent" the socio-cognitive processes at stake (through rational calculations and statistical modeling) and, on the other hand, a normative conception of – social – acceptability which, in the name of a certain "scientific" objectivity, seeks to "represent" the socio-cognitive processes at stake (through rational calculations and statistical modeling), a more situated approach to acceptance (on which we find ourselves), which seeks to "embody" the dynamics of adoption and appropriation to the work, through lived experience, experience, the subjective character of knowledge and situational interpretation.

We can draw a certain parallel between these ways of looking at acceptance (*represented vs. lived*) and the critical analysis held by the philosopher Simone Weil on the subject of psychotechnology, which was massively deployed at the beginning of the 20th Century in companies to measure the working conditions of workers:

> Since Taylor, a special branch of science has developed [...] it is called psychotechnology, which makes it possible to define the best possible psychological conditions for such and such a job, to measure fatigue, etc. This is called psychotechnics, which makes it possible to define the best possible psychological conditions for such and such a job, to measure fatigue, etc. Then industrialists, thanks to

> psychotechnics, can say that they have the proof that they don't make their workers suffer. All they have to do is invoke the authority of scientists. But psychotechnics is still imperfect. It has only just been created. And even if it were perfect, it would never reach the moral factors; for suffering in the factory consists above all in finding the long time; but it does not stop there. And no psychoanalyst will ever be able to specify to what extent a worker finds time long. It is the worker himself who can say so. (Weil 2002, p. 303)

In other words, consequentially, what is lacking in social acceptability is put into perspective in the reality of the activity and in the person's actual experience, effective and subjective, in order to give meaning and value to this device in a situation of concrete use.

From this perspective, acceptance (unlike social acceptability) is seen as an emergent process, that is, it is always socially situated and physically embodied. It is therefore not (solely or simply) the fulfillment of predefined social representations and plans (as with social acceptability). On the contrary, attitudes (favorable or unfavorable) towards technology are the product of a situation-based user experience. In other words, the issue is not so much that the technology is perceived favorably (implying that the individual has a set of favorable attitudes towards it: the case of social acceptability), but that the technologies are actually favorable to the activity (implying that they have a favorable effect on the activity: the perspective of situated acceptance) and concretely support the workflow and development of the job.

We will now determine the epistemological frameworks of reflection and action of situated acceptance.

18.2.3. *The situated acceptance approach*

Situated acceptance is defined as the testing of technology in its actual context of use, which allows for a concrete evaluation of its contributions and limitations, and thus defines its relevance to the individual's activity and projects. In this perspective, it is less a question of examining the conditions of acceptability of the technological object itself, than of looking at the conditions of acceptance of new practices (which are allowed, prevented or degraded) through the use of ICTs. We are thus looking at what technology "allows/authorizes to be done" or "obliges to be done", as well as what it "prevents from being done" or "no longer allows to be done as before" on different dimensions of the activity (Bobillier Chaumon 2016). The situated acceptance approach therefore leads to not studying the technology "*per se*" and in an isolated and decontextualized manner. It leads to analysis "within its frameworks of use, which are themselves not static, but evolve and develop over time" (Rabardel 1995, p. 19). Technology is not given *a priori*; it is constructed in action and in interaction with the individual, within the framework of tasks to be accomplished and specific situations. This approach postulates the indeterminate nature of activity in interaction with innovations, emphasizing its emergent, contingent, individual and collective nature. The methodological approach starts from the activity, the actual uses and the effects on practices, to identify the complexity of the acceptance processes.

18.2.3.1. *The four dimensions of situated acceptance*

Four major dimensions are decisive factors in the situated acceptance of technologies (Bobillier Chaumon 2016) and have been studied in various works and research programs (Medzo Mengone 2016; Cros 2017; Poupon 2017; Codreanu 2018; Crouzat 2019; Poidi 2019; Barville 2020). These are the dimensions:

1) *individual (or personal)*, relating to the employee's own activities and in particular, on the effect of technology as a factor of intensification or efficiency in the activity (notion of cognitive and emotional cost);

2) *organizational (or impersonal),* which involves the employees' relationships with the organization of work, including the control exercised by technology over their actions and initiatives;

3) *relational (interpersonal)*, that corresponds to collective and collaborative activities in work and to the daily functioning of work collectives, and how tools can affect the cohesion of these teams, weakening or degrading them;

4) *professional and identity (or transpersonal)*, which express the capacity of individuals to have their know-how recognized, to maintain their professional regulations, to be able to exercise their activity and their power to act and/or to develop them.

In this framework, we see that technology is not only a condition for carrying out the activity, but also a means for the development of the activity, the profession and the individual. Technology must not only be part of the existing activity in order to be accepted (implying that it is useful, usable and compatible with the existing activity system), but also be embodied in it, it must participate in the development and enhancement of this activity. This developmental vision of acceptance therefore leads to its inclusion in a "clinic of use", which is inspired by the paradigms of the clinic of activity. Maintaining this power to act through technology is proving to be a fundamental element of occupational health and well-being.

18.3. Frameworks for action and intervention to address situated acceptance

As mentioned above, situated acceptance seeks to account for the way in which technologies can constrain or, on the contrary, promote the development of the activity and the occupation in actuality, on different dimensions of the system of activity. While it may seem "convenient" to carry out such diagnoses for common and well-established technologies (such as pre-existing business software packages or collaborative environments) using various methods of observation, interviews or user surveys (Brangier and Barcenilla 2003; Benedetto-Meyer and Chevallet 2008), it is more difficult to assess such impacts for

1) emerging technologies (Loup-Escande and Burkhardt 2019) that do not exist or are still at the prototype or demonstrator stage;

2) devices that will be deployed in activities that are difficult to capture because they are dematerialized/digitalized, divided between people (virtual teams), are used remotely (nomadic, mobile, teleworking) or shared between multiple person–machine systems;

3) and deployed to an audience with specific needs (*the frail, the very elderly, people with disabilities*), which is a target of potential users of these emerging technologies. However, these audiences have legitimate difficulties in understanding these tools (issues, functionalities) and especially in participating in the design process (in the expression of needs, projection in use, etc.) and in the involvement in a very stressful cognitive and psychological process (*questionnaires, interview scales, use tests, etc.*) (Elias and Bobillier Chaumon 2019).

How can we explore the situated acceptance of technologies under such conditions? We will present different methodological approaches that seem relevant to us, to account for these potential uses and the mediated activity in the making. They all have in common, the aim of putting this future-probable activity into perspective and discussing it, in order to instruct a reflexive and collective approach (with the different stakeholders of the design and implementation project) on the current conditions of realization of the activity and especially on the possible or impossible developments of this activity.

Through these discussion mechanisms, it is also a question of empowering individuals both in the exercise of their activity and in the design process (by intervening in the choice, development and implementation of tools). These are instruments that aim to develop the capacity of individuals and groups to act on their activity and on the technological conditions/modalities for carrying it out (Bobillier Chaumon and Clot 2016). They provide a framework so that the activity – as it is carried out, with or without these technologies – can become an object of thought.

More precisely, the dialogical and reflexive process put in place calls for the analysis of activities and confrontation (individual and/or collective) with our own practices, through a co-analysis of the digital practices deployed in the situation. The aim is to enable professionals to repeatedly examine their activity and be able to objectify the tensions of the profession on the basis of actual work and conflicts of criteria.

In the end, this methodological device seeks to support individuals in their search for new possibilities, through the displaced view they take of their mediated activity and their digital uses (current and future). The ambition is also to enable them to "*uncover*" and "*update*" their ways of thinking and acting, their practices (some of which may be inappropriate, prevented, etc.), their uses (diverted, excessive, intensive, etc.) and the modes of regulation that are deployed.

The co-analysis devices that we propose to implement, in order to explore the situated acceptance of technologies in the activity, are based on comprehensive, localized and participatory analysis methodologies. They are all based on a preliminary and detailed understanding of the activity that is being carried out and a decoding of the activity system in which these behaviors take place. This activity system, according to Engeström's (1987) understanding, presupposes that any object-oriented activity (i.e. the goal sought through the accomplishment of actions) is carried out through artifacts (material, technological) and is socially inscribed in a universe of rules (which govern the unfolding of the activity and which delimit it), in a community (composed of all the actors involved in the activity), itself linked to a division of labor (each one is responsible for a local form of processing, which cannot be understood without reference to the whole) and in a community (composed of all the actors involved in the activity): it is therefore the distribution of activities among the actors of the system. This therefore implies a multiscalar analysis at several levels of this activity system: the "micro" level of individual action in work situations, the "meso" level of collective dynamics at work and the "macro" level of organizational and managerial strategies. The objective is to be able to determine the conditions, resources and requirements of the socio-organizational context in which these discussed activities will take place.

Therefore, on the basis of this intimate knowledge of the reality of the activity and the conditions of its execution, it becomes possible to deploy the appropriate methodological device for a reflection and discussion of future uses and mediated practices, in order to explore the situated acceptance of the technology.

In this perspective, different methodologies can be mobilized. They are chosen and adjusted according to the possibilities of access to the situation and to the population (*study of domestic activities vs. professional activities, nomadic vs. sedentary, dependent vs. able-bodied people*), to the types of evaluated technologies (*mature vs. emergent, for individual use vs. collective, etc.*) to the stage of development of these devices (*design, evaluation, dissemination, use phase*) and of course, to the empirical aims of these approaches (*reflective analysis vs. prospective analysis, simulation vs. feedback*). The objective of the data collected is not only to evaluate the acceptance of the technologies in the activity that is being carried out, and in the one that could be carried out, but also to enable all the stakeholders involved in these digital transformation projects (decision-makers, designers, managers) to see and understand the difficulties encountered, the needs expressed, the desired developments, the adjustments demanded by the future users on the technology and/or the activity system in action (which must also develop in order to integrate this system as well as possible).

Among these approaches, we can distinguish (1) those that will make it possible to evaluate the acceptance of technologies already in place that we would like to transform or develop, and (2) methods that aim to explore the acceptance of emerging technologies, which are still at the prototype or demonstrator stage, and are therefore very little advanced in design.

18.3.1. *Evaluating the situated acceptance of technologies already in place to support the future technological transformation project*

a) Self-confrontation

This process of formalizing professional experience, borrowed from the activity clinic (Clot *et al.* 2000), includes an analysis group, video recordings, traces of activity or, more generally, media that report on the progress of the activity. This allows for the production of filmed sequences associated with the production of simple and cross-confrontations, also filmed, and the return to the collective, as well as to the monitoring bodies (steering committee, management committee, etc.) of this work of co-analysis of the activity. The device makes it possible to better understand the complexity of the subject's work, the cognitive reasoning that accompanies his or her decisions and his or her choices of action that may be constrained by the situation for example: *why am I obliged to log on at this time of day to see my e-mails, why do I solicit my colleagues in this way, how do I favor one tool over another?*

The sense of action put into words produces an effect on others and on the subject himself and opens up new spaces for development. The participants then create a cycle between what they do, what they say about it and what they do with what they say. By showing, through this dialogue, what has been done, as well as what has not been done and what could have been done, this brings up the dilemmas of the activity and the potentialities not expressed in the activity carried out. Confrontation and controversy therefore become a possible vector for recreating modes of action, which make it possible to define collective rules for the use of tools that are more acceptable to individuals, more appropriate to their activity and more respectful of their quality criteria and business rules (and those of others). This collective and reflective experience also gives authority back to professionals over their work (by being a guarantor of the profession). They feel more legitimate and confident to discuss and negotiate with the various management bodies on the technological orientations of their work.

b) The technical object method

The second method is that of the technical object developed by Simondon in 1958 (Simondon 2015). This approach brings together a work group around a work object (a technology, a usage charter, e-mails, etc.) called a technical object. The latter is then established as a vector of communication and discussion around the work. It functions both as a federator of the given differences of a work situation and as a trigger/mediator of an analysis process within the constituted collective (Gaubert 2012). This technical object has three functions: as an analyzer of work situations through the effort to formalize the object and the way it is used, a mediator of cooperation processes and a revealer of the individual's experience and creativity. The objective is then to start from the object (emblematic of the activity), to deconstruct it in order to reconstruct it, to draw its law of functioning, to transform it and to thus be transformed. The technical object makes it possible to "*bring to the surface*" what is engaged in the work and to make the constraints visible, as well as all the professional's ingenuity. It is a question of shedding light on what makes sense to him or her, what is necessary for the accomplishment of the activity and for oneself. It leads to identifying what, in the current or future-probable use of the technical device, affects or could affect, favorably or unfavorably, the exercise of the activity, and vector of the local acceptance of technologies. The collective discussion of practices and uses also gives the possibility to (re)define the modalities of action and cooperation of the activity system, in order to integrate the technological resources in the best possible way, or to ask for adjustments of the technological project.

18.3.2. *Exploring the acceptance of emerging technologies to co-construct the creation project in/by the activity*

a) The Wizard of Oz method

The "Wizard of Oz" is a simulation experiment in which subjects are made to believe that the system which they interact with is autonomous, when in reality it is totally or partially controlled/guided by a human (Baccino *et al.* 2005). As Alevêque (2019) points out, the aim is to "ensure that a human being can imitate the functionalities of a machine that in the future will imitate his or her own" (p. 1). Concretely, the experimenter thus assumes the role of a machine that does not yet exist and tests the validity of an innovation whose technical feasibility is not yet assured (Buisine and Martin 2005). He or she is masquerading as the machine. The Wizard of Oz responds to a predictive problem by proposing to anticipate, by simulation, the nature of the relationships and adjustments required between humans and future machines that will be deployed.

In some ANR research we conducted on the acceptance of a smart fall detection technology (CIRDO), based on artificial intelligence, in the homes of the elderly, the Wizard of Oz methodology was used (Bobillier *et al.* 2016). This CIRDO system aimed, on one hand, to identify accidents related to falls (the first cause of death among the elderly) by means of "smart" cameras and microphones distributed throughout the home and, on the other hand, to automatically notify the emergency services in the event of proven difficulties (the person not responding to the system's calls following the detection of the fall).

The device deployed for the Wizard of Oz was a set of cameras and microphones distributed throughout the user's living space, which allowed the researcher – located in an adjacent room – to follow the scene on monitors.

The user was asked to reproduce scenarios of previously identified domestic activities and accidents (e.g. a person who, while napping on the couch, falls to the floor and breaks his or her nose). The researcher then pretended to simulate the services and functionalities supported by CIRDO; that is, identifying a dangerous situation (the fall) and/or a call for help ("*I am in pain, help*") and triggering the appropriate action and interaction with the user ("*I detect an abnormal situation. Do you need help?*"). These pre-recorded warning phrases were triggered by the researcher according to the elderly person's behavior and requests.

In addition to the configuration of the person–system dialogue and the identification of contact persons, this method made it possible to assess what could influence the acceptance of the future system. In particular, the fear of elderly people living alone, of making their homes more accessible and of losing control in the event of an accident: who should be notified in the event of a fall? Who should intervene and how? The subjects surveyed were apprehensive about an internal domestic risk (that of falls and accidents), as well as feared external intrusions (theft, door-to-door salesmen taking advantage of their credulity, etc.); "*The fire brigade, well, they're going to break down the door, what are they going to do? So I call them last. I don't want to break the door down!*" or "*If they break the door down and take me away, I'm not going to leave the house open.*" Experienced in the manner of a burglary, the introduction of CIRDO in the home induces representations in terms of "intrusiveness", forcing the elderly person to redefine compromises in terms of security, domestic and protection priorities, which must be taken into account in order to create acceptable conditions for the adoption of the device.

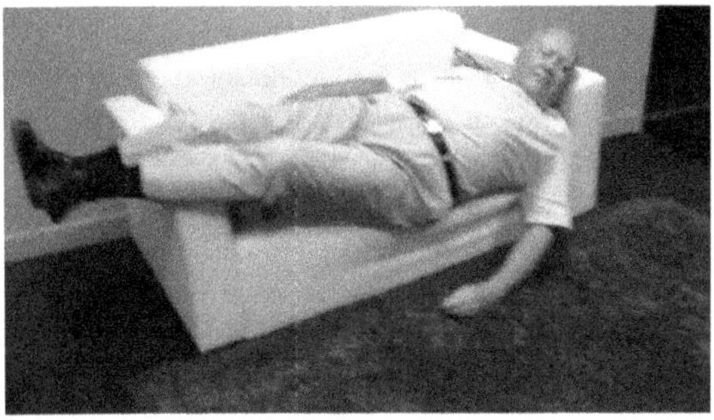

Box 18.1. *Simulation of a fall using the Wizard of Oz method*

b) The focus group method

A focus group is a collective interview method organized by a researcher who wishes to collect and bring out the expectations, fears, opinions and ideas that individuals have on a defined theme. Kitzinger *et al.* (2004) define the focus group as "a space for communication that allows one to engage, observe and analyze interactions, memories, representations [and projections, NDA] in action" (p. 238). This method makes it possible to stimulate different points of view through speech, according to a crossover between conversational and group dynamics. Subsequent analysis is thus done according to these two concomitant dimensions during the exchanges: one, pragmatic (how it is said) and the other semantic (what is said).

This method proves to be well adapted to support innovative technological solutions (Strout *et al.* 2017), especially when: (1) we wish to identify and demonstrate the gaps that exist between the design or digital transformation team on one hand, and the target users on the other; (2) we want to explain the causes of certain behaviors or opinions that may have been observed in a situation, or to understand the motivations of potential users with regard to tools; and (3) finally, we want to know the reactions of a group of users to new products or services. The methodological protocol involves two elements. First, an interview guide in the form of a scenario that indicates how the session will be conducted, as well as the aims (instructions) and times for each of these phases. Then, media (audio, video, photo, text) supposed to represent the theme, which the participants are invited to react and interact on. These media can be more or less close – symbolically, materially – to the technical object studied and/or to the reality experienced by the people involved. In any case, they must be as generic and meaningful as possible to allow each person to find elements of his or her own experience. A prior analysis is needed to ensure this is necessary.

As an example, we can mention the use of focus groups in the CIRDO project (already mentioned above) to collectively discuss and confront the way in which home carers perceived their role and their contribution in caring for the elderly, and the way in which they envisaged the evolution of their professional practices in relation to the use of this fall detection technology (Bobillier Chaumon *et al.* 2014). Different media were used in this perspective: first, a restitution in the form of a presentation of the various activities and difficulties of their profession (based on field observation) in order to discuss the conditions of exercise of the activity and to share a common approach of their professional reality; then, the broadcasting of a film accompanied by scenarios of use describing the use of CIRDO in the prevention of falls. This was followed by a debate, supervised by the facilitator, on how the system could affect their professional practices: how, at what level, with what benefits or risks; whether for them, for supported elderly people or for other professionals. These exchanges were recorded by video and transcribed in full for analysis.

The results showed that the visibility of activities created an ambivalent feeling among carers: on one hand, through the enhancement of the activity to which this technology could lead and, on the other hand, through the control that it could allow. In fact, the stakeholders believed that CIRDO contributed to the recognition of their activity by making its reality and its difficulty visible to the family or their managers. It could also protect them from abusive behaviors coming as much from their "client" (example of a household helper who was asked to do the housework naked) and from their own corporation (abusive behaviors). These are all psychosocial factors that can support or oppose the acceptance of the future CIRDO system.

Box 18.2. *Illustration of the use of the focus group method*

c) Activity simulation methods

In its general meaning, activity simulation is an approach to support transformation or technological design projects, by placing the activity and individuals at the heart of the intervention. It consists of imaginary deportation from reality by "*pretending*", that is, by being both the actor and spectator of the activity that is being done and/or will be done. It allows both "an exploration of the field of possibilities and also participates in a reduction of uncertainty" (Daniellou 2007, p. 9) insofar, as it seeks to articulate the diverse and sometimes contradictory expectations of the various stakeholders in the project. Finally, by exploring possible solutions, it aims to reconcile the development of individual activity with organizational performance. But activity simulation is more than a "simple" simulation activity. It does not only seek to test a future-probable situation in order to evaluate its relevance and appropriateness with end-users but also aims at co-constructing the conditions for carrying out this new activity, in order to better include the project in it.

Moreover, by promoting the confrontation with a possible or plausible form of reality, simulation enables the experimenting of the probable dx^2 limits of the device as well as its own limits. The individual can thus measure the impact that the device will have in the course or renewal of his or her activity: what it allows, authorizes, forces or prevents him or her from doing, as well as what it allows him or her to do better, less well or differently. These questions of efficiency and the meaning of the activity prove to be crucial in the acceptance of any innovation (Bobillier Chaumon and Clot 2016).

The ANR Vocadom research conducted on the design of an emerging voice tele-assistance technology, based on a learning machine, was aimed in particular, through the simulation method, at exploring the acceptance of the tool by vulnerable people (dependent, disabled) in the context of their socio-domestic activities (Elias *et al.* 2018). It was also a question of identifying their needs and the technical adjustments to be made by the designers. This population of vulnerable users had great difficulty in raising expectations and imagining themselves in the probable future use of an ambient technology which they knew nothing about. In order to do so, we drew inspiration from Van Belleghem's (2021) digital activity simulation method to apply it to tangible materials such as models and figurines. We opted for this approach in order to help the elderly to, on one hand, imagine their use of the Vocadom device and, on the other hand, co-construct usage scenarios using materials taken from their real lives. This miniature habitat (see Photo 18.2) was designed so that it could be transported to people's homes and personalized. It consisted of a tray with removable partitions so that the person could configure their own home. This intermediate design object allowed an overall view of the rooms and movements. Through this tool, the person replayed and explained his or her different activities at home. It thus helped people to project themselves into a possible use of Vocadom by presenting real situations where the system would intervene (e.g. *a pan that remained on the hob and the system orally alerted the person*). The person could then react to this situation and imagine the possible interaction with the device by describing what he or she feels and would like to do as actions.

Box 18.3. *Simulation of socio-domestic activity in a habitat model*

18.4. Conclusion

The different approaches mentioned in this chapter make it possible to support and/or explore the local acceptance of technologies by repositioning them in the activity that is being carried out and/or will be carried out. They also offer the advantage of not only involving the end-users in a process of co-analysis and anthropocentric co-design of the future system, but also associating all the stakeholders in the technological project (designers, decision-makers, managers). The latter can thus become aware of the direct impacts of their development choices and become aware of the reality and complexity of the activities and people for whom the technological solutions are deployed. These approaches also initiate forums for dialogue between the various project stakeholders to discuss acceptable or unacceptable uses within the framework of the users' specific activities.

These approaches are therefore a way – for individuals and groups – to give meaning (again) to the change project by placing it in a known and meaningful framework and by initiating/initiating a process of appropriation of the innovation. Indeed, it is because these different methodological devices give the possibility to influence the specification of technical objects, to express oneself on the project, to negotiate the modalities and outlines of the future activity and to be a recognized actor in the design process, that the technological project can then become acceptable to the subjects in this activity that is being constructed (Bobillier Chaumon 2016). As Daniellou (2007) pointed out, design cannot simply be a matter of inducing a prescription concerning a good way of carrying out the task. Rather, it is a matter of understanding how the likely future activity can develop and whether these possible forms are acceptable to the various protagonists, not only in relation to the situation they are currently experiencing, but also in relation to the one they will be facing in the near future.

18.5. References

Alexandre, B., Reynaud, E., Osiurak, F., and Navarro, J. (2018). Acceptance and acceptability criteria: A literature review. *Cognition, Technology & Work*, 20(1), 1–3.

Anastassova, M. (2006). L'analyse ergonomique des besoins en amont de la conception de technologies émergente : le cas de la réalité augmentée pour la formation à la maintenance automobile. PhD Thesis, Université Paris 5, Paris.

Atarodi, S., Berardi, A.M., and Toniolo, A. (2019). Le modèle d'acceptation des technologies depuis 1986 : 30 ans de développement. *Psychologie du travail et des organisations*, 23(3), 191–207.

Baccino, T., Bellino, C., and Colombi, T. (2005). *Mesure de l'utilisabilité des interfaces*. Hermes-Lavoisier, Paris.

Barville-Deromas, N. (2020). Les conditions d'acceptation des réseaux sociaux numériques d'entreprises. PhD Thesis, Université Lumière Lyon 2, Lyon.

Benedetto-Meyer, M. and Chevallet, R. (2008). *Analyser les usages des systèmes d'information et des TIC*. Anact Edition, Paris.

Bobillier Chaumon, M.E. (2016). Acceptation située des TIC dans et par l'activité : premiers étayages pour une clinique de l'usage. *Psychologie du travail et des organisations*, 22(1), 4–21.

Bobillier Chaumon, M.E. (2017). Transformation numérique et charge de travail : comprendre les liens (30), *Les Cahiers des RPS*, 28–30.

Bobillier Chaumon, M.E. and Clot, Y. (2016). Clinique de l'usage. Les artefacts technologiques comme développement de l'activité. *Activités*, 13(2), 1–7.

Bobillier Chaumon, M.E., Cuvillier, B., Durif-Bruckert, C., Cros, F., Vanhille, M., and Bekkadja, S. (2014). Concevoir une technologie pervasive pour le maintien à domicile : une démarche prospective par la prise en compte des systèmes d'activité. *Le travail humain*, 1/77, 39–62.

Bobillier Chaumon, M.E, Cuvillier B., Bekkadja, S., and Cros, F. (2016). Detecting falls at home: User-centered design of a pervasive technology. Human technology: An interdisciplinary. *Journal on Humans in ICT Environments*, 12(2), 165–192.

Brangier, E. and Barcenilla, J. (2003). *Concevoir un produit facile à utiliser. adapter les technologies à l'homme*. Edition d'organisation, Paris, France.

Brangier, E. and Bastien, J.-M.-C. (2009). L'évolution de l'ergonomie des produits informatiques : accessibilité, utilisabilité, émotionnalité et persuasivité. In *Ergonomie des produits et des services médiatisés : nouveaux territoires, nouveaux enjeux*, Valléry, G., Le Port, M.-C., and Zouinar M. (eds.). Octarès, Toulouse/Presses Universitaires de France, Paris.

Brangier, É., Dufresne, A., and Hammes-Adelé, S. (2009). Approche symbiotique de la relation humain-technologie : perspectives pour l'ergonomie informatique. *Le travail humain*, 72(4), 333–353.

Brown, S.A., Massey A.P., Montoya-Weiss M.M., and Burkman J.R. (2002). Do I really have to? User acceptance of mandated technology. *European Journal of Information Systems*, 11(4), 283–295.

Buisine, S. and Martin, J.C. (2005). Conception d'une interface multimodale à partir des comportements utilisateurs : utilisation du magicien d'Oz en conception. CPI'05 colloque francophone conception et production intégrées proceedings [Online]. Available at: http://stephanie.buisine.free.fr/publis/CPI05.pdf? [Accessed 20 April 2020].

Clot, Y. (2008). *Travail et pouvoir d'agir*. Presses Universitaires de France, Paris.

Clot, Y., Faïta, D., Fernandez, G., and Scheller, L. (2000). Entretiens en autoconfrontation croisée : une méthode en clinique de l'activité. *Pistes*, 2(1) [Online]. Available at: http://journals.openedition.org/pistes/3833 [Accessed 4 May 2020].

Codreanu, E. (2018). Les espaces numeriques de travail dans l'enseignement primaire – le role de la conception et des facteurs situationnels dans l'adoption technologique. PhD Thesis, Université Lumière Lyon 2, Lyon.

Cros, F. (2017) Les salaries ages face aux transformations digitales du travail : du role de l'environnement socio-technique dans le developpement de l'activite. PhD Thesis, Université Lumière Lyon 2, Lyon.

Crouzat, P. (2019). Collectifs de travail et mecanismes collaboratifs en ingenierie aeronautique : favoriser la collaboration inter-metiers. PhD Thesis, Université Lumière Lyon 2, Lyon.

Daniellou, F. (2007). Des fonctions de la simulation des situations de travail en ergonomie. *Activités*, 4(4–2), 77–83.

Davis, F.D. (1989). Perceived usefulness, perceived ease of use, and user acceptance of information technology. *MIS Quarterly*, 13(3), 329–340.

Dubois, M., Bobillier Chaumon, M.E., and N'Gom, L. (2015). L'acceptation des technologies : quels apports des modèles d'acceptabilité ? In *Mutualisation des ressources numériques et coordination des pratiques informationnelles*, El-Hachani, M. (ed.). Houdiard Editions, Paris.

Dupre, D., Dubois, M., and Tcherkassof, A. (2018). Rôle de l'émotion dans l'acceptabilité d'un produit : évaluation des composantes cognitive, motivationnelle et subjective. *Psychologie du travail et des organisations ?* 24(4), 313–324.

Elevêque, G. (2019). Le magicien d'Oz ou la simulation de l'interaction humain-machine. *Cultures et techniques* [Online]. Available at: http://journals.openedition.org/tc/12017.

Elias, E. and Bobillier Chaumon, M.E. (2018). Design methods for the projection of uses for vulnerable people. *Proceedings of the 20th Congress of the International Ergonomics Association (IEA 2018)*, 824, 765–773, Turin, Italy.

Elias, E., Bobillier Chaumon, ME., and Vacher, M. (2018). Methods to design home support for elders. In *Human Aspects of IT for the Aged Population. Applications in Health, Assistance, and Entertainment*, Zhou, J. and Salvendy, G. (eds). ITAP 2018. Lecture Notes in Computer Science, 10927, 276–289. Springer, Cham.

Gaubert, T. (2012). Activités and développement. In *Manuel de psychologie du travail et des organisations*, Bobillier Chaumon, M.E. and Sarnin P. (eds). DeBoek, Brussels.

Hassenzahl, M., Eckoldt, K., Diefenbach, S., Laschke, M., Lenz, E., and Kim, J. (2013). Designing moments of meaning and pleasure. Experience design and happiness. *International Journal of Design*, 7(3), 21–31.

Hsiao, C.H. and Yang, C. (2011). The intellectual development of the technology acceptance model: A co-citation analysis. *International Journal of Information Management*, 31(2), 128–136.

Kitzinger, J., Markova, I., and Kalampalikis, N. (2004). Qu'est-ce que les focus groups ? *Bulletin de psychologie*, 57(3), 237–243.

Koh, C.E., Prybutok, V.R., Ryan, S.D., Wu, Y. (2010). A model for mandatory use of software technologies: An integrative approach by applying multiple levels of abstraction of informing science. *The International Journal of an Emerging Transdiscipline*, 13.

Leduc, S., Valléry, G., Bellhari, S., and Gaston, S. (2013). *Ergonomie des interfaces logicielles : recommandations pour la conception, le développement et l'évaluation des applications*. Hermes-Lavoisier, Paris.

Loup-Escande, E. and Burkhardt, J.M. (2019). Évaluer l'utilité dans le contexte des technologies émergentes pour identifier des besoins latents : éléments issus d'une analyse des interactions en situation d'usage. *Activités*, 16(2) [Online]. Available at: http://journals.openedition.org/activites/4554 [Accessed 4 May 2020].

Medzo-Mengone, J. (2016). Consequences des technologies d'information et de communication (ntic) sur le bien-etre psychologique au travail : cas des fonctionnaires-cadres du ministere de la communication et de l'économie numerique du gabon. PhD Thesis, Université Lumière Lyon 2, Lyon.

Monéger, F., Coutarel, F. Moták, L., Chambres, P., Izaute M., and Dhome, M. (2018). L'expérience vécue et les valeurs en acte des accompagnants pour la conception d'un service de transport par navettes destinées à être autonome. *Activités*, 15(1) [Online]. Available at: http://journals.openedition.org/activites/3077 [Accessed 4 May 2020].

Neveu, C., Legrand, J.J., and Demaray, C. (2012). Teletravail : activité, risque perçu et entropie. In *Nouvelles pratiques de travail : innovations technologiques, changements organisationnels*, Vacherand-Revel J., Dubois, M., Bobillier Chaumon, M.E, Kouabenan, R. and Sarnin, P. (eds). L'Harmattan, Paris.

Nielsen, J. (1993). *Usability Engineering*. Academic Press, Boston.

Norman, D.A. (2004). *Emotional Design: Why we Love (or Hate) Everyday Things*. Basic Books, New York.

Poidi, I. (2019). Processus d'innovation collaborative dans l'entrepreneuriat : la place des artefacts dans l'activite collective en équipes de projet. PhD Thesis, Université Lumière Lyon 2, Lyon.

Poupon, L. (2017). L'acceptation de la voiture électrique : de l'acceptabilite sociale a l'acceptation située. PhD Thesis, Université Lumière Lyon 2, Lyon.

Rabardel, P. (1995). *Les hommes et les technologies*. Armand Colin, Paris.

Simondon (2015). *Du mode d'existence des objets techniques*, Aubier, Paris.

Strout, T.D., DiFazio, R.L., and Vessey, J.A. (2017). Technology-enhanced focus groups as a component of instrument development. *Nurse Researcher*, 25(1), 16–23.

Van Belleghem, L. (2021). Simulating digital activity in the making: Elements of methodology. In *Digital Transformations in the Challenge of Activity and Work: Understanding and Supporting Technological Changes*, Bobillier Chaumon, M.E (ed). ISTE Ltd, London and John Wiley & Sons, New York.

Vuori, V., Helander, N., and Okkonen, J. (2019). Digitalization in knowledge work: The dream of enhanced performance. *Cognition, Technology & Work*, 21, 237–252 [Online]. Available at: https://doi-org.proxybib-pp.cnam.fr/10.1007/s10111-018-0501-3.

Weil, S. (2002). *La condition ouvrière*. Folio, Paris.

List of Authors

Julian ALVAREZ
INSPE Lille
and
DeVisu Laboratory
Université Polytechnique
Hauts-de-France
France

Flore BARCELLINI
CRTD Laboratory
CNAM
Paris
France

Nadia BARVILLE-DEROMAS
GRePS Laboratory
Lumière University Lyon 2
France

Tahar-Hakim BENCHEKROUN
CRTD Laboratory
CNAM
Paris
France

Sandrine BERGER-DOUCE
Institut Mines-Télécom
Coactis
Jean Monnet University
Saint-Etienne
France

Maxime BESENVAL
Centre de Sociologie des
Organisations (CSO) SNRS
EHESS
France

Alexandra BIDET
CNRS
Centre Maurice Halbwachs
(CNRS-ENS-EHESS)
France

Marc-Eric BOBILLIER CHAUMON
CRTD Laboratory
CNAM
Paris
France

Flavie BONNEVIOT
PErSEUs Laboratory
University of Lorraine
and
Institut VEDECOM mobiLAB
Versailles
France

Eric BRANGIER
PErSEUs Laboratory
University of Lorraine
France

Willy BUCHMANN
CRTD Laboratory
CNAM
Paris
France

Elodie CHAMBONNIÈRE
GRePS Laboratory
Lumière University Lyon 2
France

Raluca CIOBANU
EDF Laboratory R&D
Lyon
France

Clément COLIN
PErSEUs Laboratory
University of Lorraine
and
Total SA
Paris
France

Clotilde CORON
Pantheon-Sorbonne University
Paris
France

Florence CROS
CRTD Laboratory
CNAM
Paris
France

Pauline CROUZAT
GRePS laboratory
Lumière University Lyon 2
France

Bruno CUVILLIER
GRePS laboratory
Lumière University Lyon 2
France

Tamari GAMKRELIDZE
Orange Labs
CRTD Laboratory
CNAM
Paris
France

Patrick GILBERT
Pantheon-Sorbonne University
Paris
France

Maria IANEVA
CRTD Laboratory
CNAM
Paris
France

Anne-Cécile LAFEUILLADE
CRTD Laboratory
CNAM
Paris
France

Chiara LAI
Colliers International France R&D
CRTD
CNAM
Paris
France

Émilie LOUP-ESCANDE
Centre de Recherche en Psychologie
University of Picardy Jules Verne
France

Masood MALDAR
Polytechnique Montréal
Canada

Antoine MARTIN
PErSEUs Laboratory
University of Lorraine
France

Lydia MARTIN
CRTD Laboratory
Paris
France

Irène POIDI
ELICO Laboratory
University of Lyon 2
France

Jean-Marc ROBERT
Polytechnique Montréal
Canada

Camille SAGNIER
Centre de Recherche en Psychologie
University of Picardy Jules Verne
France

Antoine TALY
CNRS
University of Paris
France

Mitra TARAGHI
Polytechnique Montréal
Canada

Jacqueline VACHERAND-REVEL
Ecole Centrale of Lyon
GRePS Laboratory
Lumière University Lyon 2
France

Gérard VALLÉRY
Centre de Recherche en Psychologie
University of Picardy Jules Verne
France

Laurent VAN BELLEGHEM
Realwork/Plein Sens
Paris
France

Emilie VAYRE
GRePS Laboratory
Lumière University Lyon 2
France

Moustafa ZOUINAR
Orange Labs
CRTD Laboratory
CNAM
Paris
France

Index

A, B, C

acceptance, 40, 57, 94, 102, 133, 137, 138, 178, 233, 238–248, 250–252
accessibility, 32, 93, 239, 248
activity analysis, 71, 72, 83
affect, 59, 60, 68
anthropocentric, 40, 252
appropriation, 24, 61, 96, 102, 113, 137, 150, 158, 177, 179, 192, 217, 218, 238, 240, 241, 252
arduousness, 48, 51, 83, 116, 123, 124, 229
artificial intelligence (AI), 4–6, 14, 31, 71, 229, 238, 248
augmented reality (AR), 4, 7
automation, 77–80, 83, 226–229
Big Data, 4, 6, 9, 71, 73, 212, 227–229
burnout, 91, 92, 227, 229, 234
change management, 116–118, 125, 150, 225, 232–234
cobots, 4, 53, 55, 122
co-configuration of work, 22
collaboration, 10, 15, 21, 64, 78–83, 93, 102, 109, 113, 148, 152, 184–187, 191, 238
  asynchronous, 22, 26, 27
collaborative
  platforms, 8, 21, 27, 137
  tools, 185, 187, 190, 192
collective
  activity, 25, 27, 130, 151, 156, 183, 185, 187, 190, 192
  work, 10, 97, 137, 155, 159, 183, 185, 190, 191
comfort, 96, 150, 173, 200, 232
commitment, 14, 21, 26, 106, 155, 156, 161, 165, 166, 213, 238
connection
  dis-, 95, 231–234
  hyper, 227
  over, 14
connectionist approach, 73
cooperation, 5, 10, 23, 60, 78–80, 83, 92, 129, 133, 137, 184–186, 218, 219, 247
coordination, 129, 130, 132, 183, 186, 191, 217
creativity, 80, 129–132, 148, 160, 192, 198, 201, 202, 204, 208, 215, 228, 229, 247

D, E, F

deep learning, 5, 73, 74
emerging technologies, 3, 54, 237
emotion (*see also* feeling), 57–62, 64, 65, 68, 113, 202, 239
emotional labor, 57–60, 64, 66–68
ethics, 76, 204

evaluation, 11, 26, 38, 63, 67, 95, 96, 134, 135, 184, 200, 206, 239, 241, 242, 246
expert system, 6, 73, 74, 78, 80
face-to-face, 11, 96, 101, 217, 222
feeling (*see also* emotion), 8–11, 14, 33, 59, 60, 65, 68, 92, 161, 250

I, K, M

immersive, 4, 6, 7, 33–35, 37, 39, 40
innovation, 4, 5, 8, 14, 23, 27, 51, 54, 101, 124, 129, 144, 149, 152, 197, 213, 229–234, 239, 240, 243, 248, 251, 252
intensification, 3, 91, 147, 243
knotworking, 25, 26, 164
machine learning, 5, 71, 73, 74, 83
meaning, 6, 13–15, 24, 25, 27, 55, 131, 151, 155, 157, 161, 165, 213, 238, 242, 250–252
media coverage of activities, 102
multi-mediation, 24, 26, 27

N, O, P

networking, 21, 23, 24, 55, 187
nomadism, 3, 11, 12, 90, 219, 244, 245
occupational health, 3, 15, 123, 124, 152, 243
older employees, 169
poly-contextuality, 24–27
power to act, 21, 23–25, 27, 112, 118, 137, 183, 216, 243
prospective ergonomics (PE), 197

R, S, T

reality
  augmented (AR), 4, 7
  virtual (VR), 7, 31, 223
recognition, 6, 9, 36, 47, 49, 59, 76, 93, 174, 178, 197, 201, 221, 238, 250
robots, 4, 5, 9, 13, 31, 45, 73, 115, 155
serious game, 57
  escape (SEG), 58–64, 66, 67
shared collaborative intentionality, 21, 26, 27
simulation, 7, 32–34, 36, 38, 39, 73, 82, 83, 203, 211, 246–251
SMEs (small and medium-sized enterprises), 31, 45, 46
social capital, 22, 23
socio-cognitive resources, 187
substitution, 77, 78, 155
supervision, 14, 108, 111
technocentric, 40, 115, 116
telework, 12, 14, 89, 129, 227, 229, 233, 234, 244

U, V, W

usability, 40, 82, 202, 208, 213, 239
utility, 76, 240
valorization, 9
virtual reality (VR), 7, 31, 223
visibility, 9, 23, 79, 111, 112, 144, 147, 159, 161, 184, 187, 190, 191, 250
well-being, 3, 4, 9, 40, 59, 147, 198, 199, 232, 238, 243
work
  activity, 23, 24, 27, 76, 79, 82, 83, 96, 147, 171, 211, 213
  -load, 93, 95, 174
  mediatized, 105
  organization, 27, 95, 184, 192, 243
  quality of, 76, 90, 91, 174, 183, 191, 238
    life at, 76, 93, 97, 171, 191, 238
  -space, 143–145, 148–152, 157, 159, 160, 222
  virtual, 212
working
  conditions, 7, 27, 50–53, 96, 119, 120, 175, 227, 229–231, 233–235, 241
  remote, 3, 10, 12, 13, 90, 92, 95–97, 112, 185–187, 191, 200, 222, 229, 231, 232, 244
  time, 91

Other titles from

ISTE

in

Innovation, Entrepreneurship and Management

2020

ANDREOSSO-O'CALLAGHAN Bernadette, DZEVER Sam, JAUSSAUD Jacques, TAYLOR Robert
Sustainable Development and Energy Transition in Europe and Asia (Innovation and Technology Set – Volume 9)

BEN SLIMANE Sonia, M'HENNI Hatem
Entrepreneurship and Development: Realities and Future Prospects (Smart Innovation Set – Volume 30)

CHOUTEAU Marianne, FOREST Joëlle, NGUYEN Céline
Innovation for Society: The P.S.I. Approach (Smart Innovation Set – Volume 28)

CORON Clotilde
Quantifying Human Resources: Uses and Analysis (Technological Changes and Human Resources Set – Volume 2)

CORON Clotilde, GILBERT Patrick
Technological Change (Technological Changes and Human Resources Set – Volume 1)

CERDIN Jean-Luc, PERETTI Jean-Marie
The Success of Apprenticeships: Views of Stakeholders on Training and Learning
(Human Resources Management Set – Volume 3)

DELCHET-COCHET Karen
Circular Economy: From Waste Reduction to Value Creation
(Economic Growth Set – Volume 2)

DIDAY Edwin, GUAN Rong, SAPORTA Gilbert, WANG Huiwen
Advances in Data Science
(Big Data, Artificial Intelligence and Data Analysis Set – Volume 4)

DOS SANTOS PAULINO Victor
Innovation Trends in the Space Industry
(Smart Innovation Set – Volume 25)

GOGLIN Christian
Emotions and Values in Equity Crowdfunding Investment Choices 1: Transdisciplinary Theoretical Approach

GUILHON Bernard
Venture Capital and the Financing of Innovation
(Innovation Between Risk and Reward Set – Volume 6)

LATOUCHE Pascal
Open Innovation: Human Set-up
(Innovation and Technology Set – Volume 10)

MAKRIDES Andreas, KARAGRIGORIOU Alex, SKIADAS Christos H.
Data Analysis and Applications 3: Computational, Classification, Financial, Statistical and Stochastic Methods
(Big Data, Artificial Intelligence and Data Analysis Set – Volume 5)
Data Analysis and Applications 4: Financial Data Analysis and Methods
(Big Data, Artificial Intelligence and Data Analysis Set – Volume 6)

MASSOTTE Pierre, CORSI Patrick
Complex Decision-Making in Economy and Finance

MEUNIER François-Xavier
Dual Innovation Systems : Concepts, Tools and Methods
(Smart Innovation Set – Volume 31)

MICHAUD Thomas
Science Fiction and Innovation Design (Innovation in Engineering and Technology Set – Volume 6)

MONINO Jean-Louis
Data Control: Major Challenge for the Digital Society
(Smart Innovation Set – Volume 29)

MORLAT Clément
Sustainable Productive System: Eco-development versus Sustainable Development
(Smart Innovation Set – Volume 26)

SAULAIS Pierre, ERMINE Jean-Louis
Knowledge Management in Innovative Companies 2: Understanding and Deploying a KM Plan within a Learning Organization
(Smart Innovation Set – Volume 27)

2019

AMENDOLA Mario, GAFFARD Jean-Luc
Disorder and Public Concern Around Globalization

BARBAROUX Pierre
Disruptive Technology and Defence Innovation Ecosystems
(Innovation in Engineering and Technology Set – Volume 5)

DOU Henri, JUILLET Alain, CLERC Philippe
Strategic Intelligence for the Future 1: A New Strategic and Operational Approach
Strategic Intelligence for the Future 2: A New Information Function Approach

FRIKHA Azza
Measurement in Marketing: Operationalization of Latent Constructs

FRIMOUSSE Soufyane
Innovation and Agility in the Digital Age
(Human Resources Management Set – Volume 2)

GAY Claudine, SZOSTAK Bérangère L.
Innovation and Creativity in SMEs: Challenges, Evolutions and Prospects
(Smart Innovation Set – Volume 21)

GORIA Stéphane, HUMBERT Pierre, ROUSSEL Benoît
Information, Knowledge and Agile Creativity
(Smart Innovation Set – Volume 22)

HELLER David
Investment Decision-making Using Optional Models
(Economic Growth Set – Volume 2)

HELLER David, DE CHADIRAC Sylvain, HALAOUI Lana, JOUVET Camille
The Emergence of Start-ups
(Economic Growth Set – Volume 1)

HÉRAUD Jean-Alain, KERR Fiona, BURGER-HELMCHEN Thierry
Creative Management of Complex Systems
(Smart Innovation Set – Volume 19)

LATOUCHE Pascal
Open Innovation: Corporate Incubator
(Innovation and Technology Set – Volume 7)

LEHMANN Paul-Jacques
The Future of the Euro Currency

LEIGNEL Jean-Louis, MÉNAGER Emmanuel, YABLONSKY Serge
Sustainable Enterprise Performance: A Comprehensive Evaluation Method

LIÈVRE Pascal, AUBRY Monique, GAREL Gilles
Management of Extreme Situations: From Polar Expeditions to Exploration-Oriented Organizations

MILLOT Michel
Embarrassment of Product Choices 2: Towards a Society of Well-being

N'GOALA Gilles, PEZ-PÉRARD Virginie, PRIM-ALLAZ Isabelle
Augmented Customer Strategy: CRM in the Digital Age

NIKOLOVA Blagovesta
The RRI Challenge: Responsibilization in a State of Tension with Market Regulation
(Innovation and Responsibility Set – Volume 3)

PELLEGRIN-BOUCHER Estelle, ROY Pierre
Innovation in the Cultural and Creative Industries
(Innovation and Technology Set – Volume 8)

PRIOLON Joël
Financial Markets for Commodities

QUINIOU Matthieu
Blockchain: The Advent of Disintermediation

RAVIX Joël-Thomas, DESCHAMPS Marc
Innovation and Industrial Policies
(Innovation between Risk and Reward Set – Volume 5)

ROGER Alain, VINOT Didier
Skills Management: New Applications, New Questions
(Human Resources Management Set – Volume 1)

SAULAIS Pierre, ERMINE Jean-Louis
Knowledge Management in Innovative Companies 1: Understanding and Deploying a KM Plan within a Learning Organization
(Smart Innovation Set – Volume 23)

SERVAJEAN-HILST Romaric
Co-innovation Dynamics: The Management of Client-Supplier Interactions for Open Innovation
(Smart Innovation Set – Volume 20)

SKIADAS Christos H., BOZEMAN James R.
Data Analysis and Applications 1: Clustering and Regression, Modeling-estimating, Forecasting and Data Mining
(Big Data, Artificial Intelligence and Data Analysis Set – Volume 2)
Data Analysis and Applications 2: Utilization of Results in Europe and Other Topics
(Big Data, Artificial Intelligence and Data Analysis Set – Volume 3)

UZUNIDIS Dimitri
Systemic Innovation: Entrepreneurial Strategies and Market Dynamics

VIGEZZI Michel
World Industrialization: Shared Inventions, Competitive Innovations and Social Dynamics
(Smart Innovation Set – Volume 24)

2018

BURKHARDT Kirsten
Private Equity Firms: Their Role in the Formation of Strategic Alliances

CALLENS Stéphane
Creative Globalization
(Smart Innovation Set – Volume 16)

CASADELLA Vanessa
Innovation Systems in Emerging Economies: MINT – Mexico, Indonesia, Nigeria, Turkey
(Smart Innovation Set – Volume 18)

CHOUTEAU Marianne, FOREST Joëlle, NGUYEN Céline
Science, Technology and Innovation Culture
(Innovation in Engineering and Technology Set – Volume 3)

CORLOSQUET-HABART Marine, JANSSEN Jacques
Big Data for Insurance Companies
(Big Data, Artificial Intelligence and Data Analysis Set – Volume 1)

CROS Françoise
Innovation and Society
(Smart Innovation Set – Volume 15)

DEBREF Romain
Environmental Innovation and Ecodesign: Certainties and Controversies
(Smart Innovation Set – Volume 17)

DOMINGUEZ Noémie
SME Internationalization Strategies: Innovation to Conquer New Markets

ERMINE Jean-Louis
Knowledge Management: The Creative Loop
(Innovation and Technology Set – Volume 5)

GILBERT Patrick, BOBADILLA Natalia, GASTALDI Lise,
LE BOULAIRE Martine, LELEBINA Olga
Innovation, Research and Development Management

IBRAHIMI Mohammed
Mergers & Acquisitions: Theory, Strategy, Finance

LEMAÎTRE Denis
Training Engineers for Innovation

LÉVY Aldo, BEN BOUHENI Faten, AMMI Chantal
Financial Management: USGAAP and IFRS Standards
(Innovation and Technology Set – Volume 6)

MILLOT Michel
Embarrassment of Product Choices 1: How to Consume Differently

PANSERA Mario, OWEN Richard
Innovation and Development: The Politics at the Bottom of the Pyramid
(Innovation and Responsibility Set – Volume 2)

RICHEZ Yves
Corporate Talent Detection and Development

SACHETTI Philippe, ZUPPINGER Thibaud
New Technologies and Branding
(Innovation and Technology Set – Volume 4)

SAMIER Henri
Intuition, Creativity, Innovation

TEMPLE Ludovic, COMPAORÉ SAWADOGO Eveline M.F.W.
Innovation Processes in Agro-Ecological Transitions in Developing Countries
(Innovation in Engineering and Technology Set – Volume 2)

UZUNIDIS Dimitri
Collective Innovation Processes: Principles and Practices
(Innovation in Engineering and Technology Set – Volume 4)

VAN HOOREBEKE Delphine
The Management of Living Beings or Emo-management

2017

AÏT-EL-HADJ Smaïl
The Ongoing Technological System
(Smart Innovation Set – Volume 11)

BAUDRY Marc, DUMONT Béatrice
Patents: Prompting or Restricting Innovation?
(Smart Innovation Set – Volume 12)

BÉRARD Céline, TEYSSIER Christine
Risk Management: Lever for SME Development and Stakeholder Value Creation

CHALENÇON Ludivine
Location Strategies and Value Creation of International Mergers and Acquisitions

CHAUVEL Danièle, BORZILLO Stefano
The Innovative Company: An Ill-defined Object
(Innovation between Risk and Reward Set – Volume 1)

CORSI Patrick
Going Past Limits To Growth

D'ANDRIA Aude, GABARRET Inés
Building 21st Century Entrepreneurship
(Innovation and Technology Set – Volume 2)

DAIDJ Nabyla
Cooperation, Coopetition and Innovation
(Innovation and Technology Set – Volume 3)

FERNEZ-WALCH Sandrine
The Multiple Facets of Innovation Project Management
(Innovation between Risk and Reward Set – Volume 4)

FOREST Joëlle
Creative Rationality and Innovation
(Smart Innovation Set – Volume 14)

GUILHON Bernard
Innovation and Production Ecosystems
(Innovation between Risk and Reward Set – Volume 2)

HAMMOUDI Abdelhakim, DAIDJ Nabyla
Game Theory Approach to Managerial Strategies and Value Creation
(Diverse and Global Perspectives on Value Creation Set – Volume 3)

LALLEMENT Rémi
Intellectual Property and Innovation Protection: New Practices and New Policy Issues
(Innovation between Risk and Reward Set – Volume 3)

LAPERCHE Blandine
Enterprise Knowledge Capital
(Smart Innovation Set – Volume 13)

LEBERT Didier, EL YOUNSI Hafida
International Specialization Dynamics
(Smart Innovation Set – Volume 9)

MAESSCHALCK Marc
Reflexive Governance for Research and Innovative Knowledge
(Responsible Research and Innovation Set – Volume 6)

MASSOTTE Pierre
Ethics in Social Networking and Business 1: Theory, Practice and Current Recommendations
Ethics in Social Networking and Business 2: The Future and Changing Paradigms

MASSOTTE Pierre, CORSI Patrick
Smart Decisions in Complex Systems

MEDINA Mercedes, HERRERO Mónica, URGELLÉS Alicia
Current and Emerging Issues in the Audiovisual Industry
(Diverse and Global Perspectives on Value Creation Set – Volume 1)

MICHAUD Thomas
Innovation, Between Science and Science Fiction
(Smart Innovation Set – Volume 10)

PELLÉ Sophie
Business, Innovation and Responsibility
(Responsible Research and Innovation Set – Volume 7)

SAVIGNAC Emmanuelle
The Gamification of Work: The Use of Games in the Workplace

SUGAHARA Satoshi, DAIDJ Nabyla, USHIO Sumitaka
Value Creation in Management Accounting and Strategic Management: An Integrated Approach
(Diverse and Global Perspectives on Value Creation Set –Volume 2)

UZUNIDIS Dimitri, SAULAIS Pierre
Innovation Engines: Entrepreneurs and Enterprises in a Turbulent World
(Innovation in Engineering and Technology Set – Volume 1)

2016

BARBAROUX Pierre, ATTOUR Amel, SCHENK Eric
Knowledge Management and Innovation
(Smart Innovation Set – Volume 6)

BEN BOUHENI Faten, AMMI Chantal, LEVY Aldo
Banking Governance, Performance And Risk-Taking: Conventional Banks Vs Islamic Banks

BOUTILLIER Sophie, CARRÉ Denis, LEVRATTO Nadine
Entrepreneurial Ecosystems (Smart Innovation Set – Volume 2)

BOUTILLIER Sophie, UZUNIDIS Dimitri
The Entrepreneur (Smart Innovation Set – Volume 8)

BOUVARD Patricia, SUZANNE Hervé
Collective Intelligence Development in Business

GALLAUD Delphine, LAPERCHE Blandine
Circular Economy, Industrial Ecology and Short Supply Chains (Smart Innovation Set – Volume 4)

GUERRIER Claudine
Security and Privacy in the Digital Era (Innovation and Technology Set – Volume 1)

MEGHOUAR Hicham
Corporate Takeover Targets

MONINO Jean-Louis, SEDKAOUI Soraya
Big Data, Open Data and Data Development (Smart Innovation Set – Volume 3)

MOREL Laure, LE ROUX Serge
Fab Labs: Innovative User (Smart Innovation Set – Volume 5)

PICARD Fabienne, TANGUY Corinne
Innovations and Techno-ecological Transition (Smart Innovation Set – Volume 7)

2015

CASADELLA Vanessa, LIU Zeting, DIMITRI Uzunidis
Innovation Capabilities and Economic Development in Open Economies (Smart Innovation Set – Volume 1)

CORSI Patrick, MORIN Dominique
Sequencing Apple's DNA

CORSI Patrick, NEAU Erwan
Innovation Capability Maturity Model

FAIVRE-TAVIGNOT Bénédicte
Social Business and Base of the Pyramid

GODÉ Cécile
Team Coordination in Extreme Environments

MAILLARD Pierre
Competitive Quality and Innovation

MASSOTTE Pierre, CORSI Patrick
Operationalizing Sustainability

MASSOTTE Pierre, CORSI Patrick
Sustainability Calling

2014

DUBÉ Jean, LEGROS Diègo
Spatial Econometrics Using Microdata

LESCA Humbert, LESCA Nicolas
Strategic Decisions and Weak Signals

2013

HABART-CORLOSQUET Marine, JANSSEN Jacques, MANCA Raimondo
VaR Methodology for Non-Gaussian Finance

2012

DAL PONT Jean-Pierre
Process Engineering and Industrial Management

MAILLARD Pierre
Competitive Quality Strategies

POMEROL Jean-Charles
Decision-Making and Action

SZYLAR Christian
UCITS Handbook

2011

LESCA Nicolas
Environmental Scanning and Sustainable Development

LESCA Nicolas, LESCA Humbert
Weak Signals for Strategic Intelligence: Anticipation Tool for Managers

MERCIER-LAURENT Eunika
Innovation Ecosystems

2010

SZYLAR Christian
Risk Management under UCITS III/IV

2009

COHEN Corine
Business Intelligence

ZANINETTI Jean-Marc
Sustainable Development in the USA

2008

CORSI Patrick, DULIEU Mike
The Marketing of Technology Intensive Products and Services

DZEVER Sam, JAUSSAUD Jacques, ANDREOSSO Bernadette
Evolving Corporate Structures and Cultures in Asia: Impact of Globalization

2007

AMMI Chantal
Global Consumer Behavior

2006

BOUGHZALA Imed, ERMINE Jean-Louis
Trends in Enterprise Knowledge Management

CORSI Patrick *et al.*
Innovation Engineering: the Power of Intangible Networks